T0319926

WEYERHAEUSER ENVIRONMENTAL CLASSICS

Paul S. Sutter, Editor

WEYERHAEUSER ENVIRONMENTAL CLASSICS
Paul S. Sutter, Editor

Weyerhaeuser Environmental Classics are reprinted editions of key works that explore human relationships with natural environments in all their variety and complexity. Drawn from many disciplines, they examine how natural systems affect human communities, how people affect the environments of which they are a part, and how different cultural conceptions of nature powerfully shape our sense of the world around us. These are books about the environment that continue to offer profound insights about the human place in nature.

Making Climate Change History: Primary Sources from Global Warming's Past, edited by Joshua P. Howe

Nuclear Reactions: Documenting American Encounters with Nuclear Energy, edited by James W. Feldman

The Wilderness Writings of Howard Zahniser, edited by Mark Harvey

The Environmental Moment: 1968–1972, edited by David Stradling

Reel Nature: America's Romance with Wildlife on Film, by Gregg Mitman

DDT, Silent Spring, *and the Rise of Environmentalism*, edited by Thomas R. Dunlap

Conservation in the Progressive Era: Classic Texts, edited by David Stradling

Man and Nature: Or, Physical Geography as Modified by Human Action, by George Perkins Marsh

A Symbol of Wilderness: Echo Park and the American Conservation Movement, by Mark W. T. Harvey

Tutira: The Story of a New Zealand Sheep Station, by Herbert Guthrie-Smith

Mountain Gloom and Mountain Glory: The Development of the Aesthetics of the Infinite, by Marjorie Hope Nicolson

The Great Columbia Plain: A Historical Geography, 1805–1910, by Donald W. Meinig

Weyerhaeuser Environmental Classics is a subseries within Weyerhaeuser Environmental Books, under the general editorship of Paul S. Sutter. A complete listing of the series appears at the end of this book.

NUCLEAR REACTIONS

Documenting American Encounters with Nuclear Energy

EDITED BY
JAMES W. FELDMAN

UNIVERSITY OF WASHINGTON PRESS
Seattle and London

Nuclear Reactions is published with the assistance of a grant from the Weyerhaeuser Environmental Books Endowment, established by the Weyerhaeuser Company Foundation, members of the Weyerhaeuser family, and Janet and Jack Creighton.

Printed and bound in the United States of America
Design by Thomas Eykemans
Composed in OFL Sorts Mill Goudy, typeface designed by Barry Schwartz
21 20 19 18 17 5 4 3 2 1

UNIVERSITY OF WASHINGTON PRESS
www.washington.edu/uwpress

Cataloging information is on file with the Library of Congress
ISBN 978-0-295-99962-3

The paper used in this publication is acid-free and meets the minimum requirements of American National Standard for Information Sciences—Permanence of Paper for Printed Library Materials, ANSI Z39.48–1984. ∞

For
Chris, Sam, and Ben

CONTENTS

Foreword: Postwar America's Nuclear Paradox, by Paul S. Sutter . . . . *xiii*
Acknowledgments . . . . *xvii*

Introduction: Nature and the Nuclear Consensus in Postwar America 3

PART 1 FIRST REACTIONS 21

Leslie Groves, Report on the Trinity Test, 1945 23

Harry S. Truman, White House Statement on the Bombing of Hiroshima, 1945 28

Nagasaki Mushroom Cloud, 1945 32

Joseph H. Willits, "Social Adjustments to Atomic Energy," 1946 34

Headline Comics, *Atomic Man*, 1946 39

Arthur H. Compton, "The Atomic Crusade and Its Social Implications," 1947 41

H. M. Parker, "Speculations on Long-Range Waste Disposal Hazards," 1948 45

General Advisory Committee Reports on Building the H-Bomb, 1949 49

Lewis L. Strauss to Harry S. Truman, 1949 54

PART 2 BUILDING CONSENSUS 57

"National Security Council Resolution 68," 1950 59

Federal Civil Defense Administration, *This Is Civil Defense*, 1951 . . 65

Federal Civil Defense Administration, *Women in Civil Defense*, 1952 69

Dwight D. Eisenhower, "Address before the General Assembly of the United Nations on Peaceful Uses of Atomic Energy," 1953 . . 74

Union Carbide and Carbon Corporation, "What does Atomic Energy really mean to you?" 1953 79

Lewis L. Strauss, "My Faith in the Atomic Future," 1955 81

Heinz Haber, *The Walt Disney Story of Our Friend the Atom*, 1956 . . 86

Bureau of Public Roads, *A Preliminary Report on Highway Needs for Civil Defense*, 1956 93

Walter Reuther, *Atoms for Peace: A Separate Opinion*, 1956 96

PART 3 CHALLENGING CONSENSUS 101

Bertrand Russell and Albert Einstein, "The Russell-Einstein Manifesto," 1955 103

Roger Revelle and Milner B. Schaefer, "General Considerations Concerning the Ocean as a Receptacle for Artificially Radioactive Materials," 1957 107

Atomic Energy Commission, *Atomic Tests in Nevada*, 1957 111

National Committee for a Sane Nuclear Policy, "We Are Facing a Danger Unlike Any Danger That Has Ever Existed," 1957 114

Atomic Energy Commission, *Atoms for Peace U.S.A.*, 1958 118

Barry Commoner, "The Fallout Problem," 1958 123

Edward Teller, "The Plowshare Program," 1959 128

Office of Civil Defense and Mobilization, Fallout Maps, 1959 132

Herman Kahn and H. H. Mitchell, The Postattack Environment, 1961 134

Margaret Mead, "Are Shelters the Answer?" 1961 140

Women Strike for Peace Milk Campaign, 1961 144

Atomic Energy Commission, *Annual Report*, 1962 149

John F. Kennedy, "Commencement Address at American University," 1963 152

David E. Lilienthal, *Change, Hope, and the Bomb*, 1963 155

John F. Kennedy, "Address to the American People on the Nuclear Test Ban Treaty," 1963 160

PART 4 CONFRONTING PARADOX 163

Glenn T. Seaborg, *Environmental Effects of Producing Electric Power*, 1969 165

Minnesota Environmental Control Citizens Association, "A Nuclear Energy Gamble," ca. 1969 171

Lenore Marshall, "The Nuclear Sword of Damocles," 1971 175

Calvert Cliffs' Coordinating Committee, Inc., v. United States Atomic Energy Commission, 1971 179

William R. Gould, "The State of the Atomic Industry," 1974 183

Committee on the Present Danger, "Common Sense and the Common Danger," 1976 188

Ralph W. Deuster, "R_x for the 'Back' of the Cycle," 1976 192

Leonard Rifas, *All-Atomic Comics*, 1976 196

David N. Merrill, *Nuclear Siting and Licensing Process*, 1978 198

Helen Caldicott, *Nuclear Madness*, 1978 200

Abalone Alliance, "Declaration of Nuclear Resistance," 1978 . . 204

Report of the President's Commission on the Accident at Three Mile Island, 1979 207

Gloria Gregerson, Radiation Exposure and Compensation, 1981 . . 211

PART 5 RENEWAL 215

David E. Lilienthal, *Atomic Energy: A New Start*, 1980 217

Ronald Reagan, "Address to Members of the British Parliament," 1982 221

Nuclear Waste Policy Act of 1982 225

Jonathan Schell, *The Fate of the Earth*, 1982 230

Ronald Reagan, "Address to the Nation on Defense and National Security," 1983 234

Carl Sagan, "The Nuclear Winter," 1983 238

Office of Technology Assessment, *Nuclear Power in an Age of Uncertainty*, 1984 246

Campaign for a Nuclear Free Future, *Disarmament Begins at Home*, ca. 1984 251

Bernard Lown, "A Prescription for Hope," 1985 255

Elizabeth Macias, *High-Level Nuclear Waste Issues*, 1987 259

Ronald Reagan, "Address to the 42nd Session of the United Nations," 1987 263

Editors of the *Bulletin of the Atomic Scientists*, "A New Era," 1991 . . 267

EPILOGUE THE NUCLEAR PRESENT 271

David Albright, Kathryn Buehler, and Holly Higgins, "Bin Laden and the Bomb," 2002 273

Allison M. Macfarlane, "Yucca Mountain and High-Level Nuclear Waste Disposal," 2006 277

Oregon Department of Energy, *Hanford Cleanup: The First 20 Years*, 2009 282

Mark Z. Jacobson, "Nuclear Power Is Too Risky," 2010 288

President's Blue Ribbon Commission on America's Nuclear Future, *Report to the Secretary of Energy*, 2012 291

Nuclear Energy Institute, "Nuclear Energy: Powering America's Future," 2013 296

Ken Caldeira, Kerry Emanuel, James Hansen, and Tom Wigley, "To Those Influencing Environmental Policy but Opposed to Nuclear Power," 2013 300

Latuff Cartoons, Fukushima Cartoon, 2014 303

John Asafu-Adjaye et al., "An Ecomodernist Manifesto," 2015 . . 305

Index 311

FOREWORD

Postwar America's Nuclear Paradox

PAUL S. SUTTER

Some time in the near future, the International Commission on Stratigraphy (ICS) will make a momentous decision: whether or not to designate a new geological epoch called the "Anthropocene." This decision, though a complex and technical one, hinges on whether humans have become such a dominant force on this planet that they will leave an unambiguous signature in the earth's strata, one that would constitute a clear dividing line in the planet's geological history. A collection of scholars known as the Anthropocene Working Group has been tasked with making an initial recommendation to the ICS and with deciding when to date the beginning of the Anthropocene. Although there are several viable contenders, including the onset of the Industrial Revolution and the widespread combustion of fossil fuels that came with it, these scholars are increasingly focused on a more recent beginning point: July 16, 1945, the day that Americans exploded the first atomic bomb in Alamogordo, New Mexico. As a result of that detonation, known as the Trinity Test, as well as the two bombs that the Americans subsequently dropped on Japan and the hundreds of aboveground tests that followed over the next two decades, radioactive debris has settled in sediments all across the planet in ways that will be stratigraphically recognizable for a long time. With nuclear radiation, humankind has etched its signature across the earth's surface.

Whether or not the ICS recognizes the Anthropocene as a new epoch with a punctuated beginning in July 1945, it is clear that the dawn of the Atomic Age ushered in a new day in humankind's collective environmental history. The achievement of controlled nuclear reactions capable of generating unfathomable power was a crowning success of modern American science and technology, one that J. Robert Oppenheimer (lead scientist on the Manhattan Project) famously characterized as "technically sweet." American and international scientists and engineers had unlocked an elemental source of unbounded energy that promised to change human life on earth. To use an apt metaphor, humankind had "harnessed the atom." At the same time, however, many people came to rue the realization of nuclear power as a feat that threatened to destroy the fabric of life on earth. The Promethean act of unleashing the energy of the atom symbolized both unprecedented human mastery of nature and the considerable dangers that came with such comprehensive environmental power and control. Oppenheimer apparently later recognized this too, in an ominous invocation of a verse from the Bhagavad Gita: "I am become Death, destroyer of worlds." The very notion of human progress pivoted as a result of this earth-shattering achievement.

In *Nuclear Reactions*, an essential and timely collection of primary source documents on the history of Americans' reactions to nuclear energy, volume editor James W. Feldman helps us to see that the Atomic Age was indeed a profound environmental watershed. This is why this book finds such a comfortable home in the Weyerhaeuser Environmental Classics series. But Feldman also wants us to consider the myriad other ways in which the arrival of nuclear energy shaped postwar American history. If the harnessing of the atom represented a new chapter in the history of human-environmental relationships, it also ushered in a novel moment in the political, social, economic, and cultural history of the United States. Although the environmental frame is thus a necessary and overarching one for understanding American reactions to nuclear energy after World War II, it is not a sufficient one. Indeed, the documents that Feldman has skillfully assembled in *Nuclear Reactions* show us that nuclear energy not only occupied a central place in postwar American history, but that in the Atomic Age mainstream U.S. history was necessarily environmental history. This collection of classic documents will be as useful to modern U.S. history courses as it will be to environmental history and environmental studies courses.

Feldman has adeptly organized these documents around what scholars have called "the nuclear paradox": that nuclear energy, in its various forms, has been the subject of both technological optimism and apocalyptic pessimism. Americans have seen nuclear energy as both the greatest hope for humankind and our gravest threat. Few scientific and technological achievements, Feldman suggests, have raised so many profound questions about Americans' relations with the natural world, the shape of civil society, and the very meanings of progress. Where some see commercial nuclear power as a vital source of clean energy, others insist that the potential for nuclear accidents and the unresolved problem of radioactive waste make nuclear power anything but clean. Where some see a nuclear-powered future as essential for democratic affluence and the continued flourishing of our consumer economy, others fear that the nature of nuclear power generation, which requires centralized control and tight security, will embolden technocratic elites to remove energy decisions from democratic control. Where many have seen a strong nuclear arsenal as essential to American strength in the world and the protection of the American way of life, others (particularly in the post–Cold War world) have become deeply anxious about how rogue nations or nonstate actors might use nuclear attacks to destabilize the world order and undermine U.S. power. Where some have celebrated nuclear energy as a culminating scientific and engineering triumph, others have used the myriad threats posed by it to fundamentally question whether continuing advances in science and technology constitute progress at all. And although Americans have debated nuclear energy, as they have many other important issues that have become the topics of readers like this one, they have also been deeply confused and conflicted about its promise and perils. Even Oppenheimer, the so-called father of the atomic bomb, could not quite plumb what he had done by unleashing the power of the atom. Most Americans have, in their reactions to nuclear power, experienced this same kind of deep ambivalence.

Nothing better illustrates the paradox at the center of *Nuclear Reactions* than the historical relationship of American environmentalism to nuclear energy. On one level, nuclear technology was an essential catalyst to the rise of postwar U.S. environmentalism, a movement that mixed traditional concerns about environmental protection with fears about novel threats to human health. On balance, and for most of the postwar period, environmentalists have been among the most consistent critics of nuclear energy in all of its various forms. This collection makes that clear. But, ironically, the very nature of radiation as an environmental threat also allowed postwar ecologists to study and conceptualize environmental systems and energy flows in powerful new ways. Nuclear energy may have threatened the fabric of life on earth, but it also revolutionized how scientists understood that fabric. Moreover, in recent decades, as human-induced climate change has raised the specter of profound environmental instability, some environmentalists have come to believe that nuclear power is necessary to weaning us from our destructive dependency on fossil fuels. Even with its risks, these advocates suggest, only nuclear power has the capacity to meet the world's present and future energy needs while substantially reducing greenhouse gas emissions. To put it more bluntly, while nuclear energy may have ushered in the Anthropocene, it may also be a critical technology for mitigating and managing the Anthropocene's worst effects.

The great virtue of *Nuclear Reactions* is how it uses a thoughtfully curated set of primary source documents to, as Feldman puts it, "connect the nuclear past to the nuclear future." Nuclear energy is not going away, and we will need to make reasoned decisions about whether it can, and how it should, fit into our twenty-first-century world. These are not decisions that we can make only on narrow technical grounds, guided by a naïve faith in the inevitability of technological progress and the capacity of rational actors to efficiently manage nuclear systems. Nor are these decisions that we ought to cede to technical experts. The history of nuclear energy is filled with too many cases of human failure, and of human and environmental harm, to sustain such uncritical trust in expertise or faith in cornucopian progress. But neither can we reject nuclear energy out of hand, insisting that unloosing the power of the atom constituted a fundamental sin against nature. Instead, these documents raise questions about responsible citizenship and the need for Americans to critically engage with nuclear energy in all of its historical complexity. Invariably, the future of nuclear energy will be built upon this history of nuclear reactions.

ACKNOWLEDGMENTS

This project started as a sidelight. As I began the research for a book on the history of radioactive waste management in the United States, I developed a new appreciation for the ways that nuclear energy has touched just about every aspect of American life. The more I learned about the narrow topic of waste management, the more I recognized the breadth of material I would need to understand to do the topic justice. There is no better way to learn a subject than to teach it, so I created a course that explores the many facets of American encounters with nuclear energy from the perspective of environmental history. I first developed this collection of documents for use in that course.

As I delved more deeply into these documents, the more fascinated I became. Movies, comic books, private letters, public speeches—so many interesting and powerful documents to choose from! When people discussed nuclear energy, they grappled with big, sweeping questions in such meaningful ways. While this collection focuses on the themes of relations with nature, the shape of civil society, and definitions of progress, I could easily have chosen other issues: social and environmental justice; the use of science; diplomacy and international relations. Even after narrowing the documents to a set of three themes, I could not include all of the relevant topics and perspectives. For example, these documents do not address the mining of uranium in the American Southwest and the bitter legacy this left for both landscapes and for communities. Indeed, a different editor would look at the same general subject and choose entirely different documents. It is this incredible diversity that makes the subject of American encounters with nuclear energy so rich.

When I first pitched the idea to Bill Cronon and Marianne Keddington-Lang at the University of Washington Press, they offered immediate support and encouragement. So did Paul Sutter and Regan Huff when they picked up the editorial baton of the Weyerhaeuser Environmental

Books series. Their advice and guidance helped keep the project on track and reined in my ambitions for a wider, more diffuse collection.

As I embarked on the search for documents, I turned to friends and colleagues—and sometimes strangers—for advice and for recommendations in their areas of expertise. Almost without fail, they responded, and for this many thanks to Michael Egan, Jake Hamblin, Scott Kirsch, Paul Rubinson, Jennifer Thomson, Marsha Weisiger, and Tom Wellock. I am sure there are others whom I forget to thank. Tom and Jennifer also had extended conversations with me on the general shape of the reader and the table of contents, and Tom's thoughtful commentary in the review process greatly improved the collection. David Stradling helped me understand the process involved in compiling a documentary reader. Chris Wells and David Bernstein twice read drafts of the introduction, providing invaluable comments and critiques.

A number of individuals, archives, and institutions allowed me to reproduce their work free of charge, including the American Philosophical Association, Union Carbide Corporation, Reader's Digest, United Auto Workers, the Bertrand Russell Peace Foundation, the Swarthmore College Peace Collection, the Wisconsin Historical Society, the Minnesota Historical Society, the Wilderness Society, the J. Willard Marriott Library at the University of Utah, Leonard Rifas, the Abalone Alliance, the Massachusetts Medical Foundation, Mark Z. Jacobson, the Nuclear Energy Institute, and Michael Shellenberger. This collection would not be have been possible without their generosity.

At the University of Wisconsin Oshkosh, students unknowingly acted as guinea pigs as I tested different documents in the classroom. Their suffering, I hope, will make this book more appealing to students who read these documents in other classrooms. The University of Wisconsin Oshkosh Office of Grants and Faculty Development provided funds to purchase reprint permissions and grants to help me develop the project. My student Andrea Vitale painstakingly cross-read the majority of the documents against the originals. Jeri Zelke, Kim Bullington, Jessi Stamn, and Kayla Much provided much appreciated office and administrative support. Other colleagues provided hallway—and carpool—encouragement and constant good cheer.

Like many of the authors whose work I studied in selecting documents, contemplating American encounters with nuclear energy makes me think of the future and wonder about how we will deal with the choices that lie ahead. With the hope that an understanding of the histories described in this book provides guidance—or at least helps us ask good questions—I dedicate this book to my family.

NUCLEAR REACTIONS

INTRODUCTION

Nature and the Nuclear Consensus in Postwar America

In the spring of 2011 the prospects for commercial nuclear power in the United States looked bright. Industry groups reported that public support for nuclear power had reached 74 percent, the highest recorded figure in more than three decades. A Georgia utility company had received permitting approval from the Nuclear Regulatory Commission and had started construction on the first new generating stations to be built since the accident at Three Mile Island in 1979. Nuclear power seemed like a safe, clean, low-carbon response to the threat of climate change. Even many environmentalists were calling for the industry's rebirth. But on March 11, 2011, the earthquake, tsunami, and meltdown at Japan's Fukushima Daiichi power plant once again changed public support for nuclear energy. A forty-foot wave surged over the plant's protective sea wall, disabling its electricity and cooling systems and leading to meltdowns in three reactors and plumes of radioactive smoke from fires in the spent fuel stores. In the following months and years, Americans watched with trepidation as a radiation plume from Fukushima slowly drifted across the Pacific. An online news source announced: "Don't Panic! Fukushima Radiation Just Hit the West Coast." One observer called Fukushima a "'focusing event'—a crisis that generates massive media and public attention and ripple effects well beyond the disaster itself." Popular support for nuclear power promptly dropped as people feared once again the potential for disaster.[1]

American perceptions of other elements of nuclear energy have been just as difficult to pin down. With the demise of the Soviet Union in the 1990s, the United States reigns as the world's foremost military power, commanding a nuclear arsenal far surpassing that of any other nation. But no one quite knows how nuclear weapons fit into a security strategy for a world with only one superpower. The global War on Terror has focused on an enemy that, it seems, might be all too willing to use a nuclear weapon, as the Cold War rules of deterrence predicated on a relative nuclear balance between two superpowers no longer apply. The instability and nuclear ambitions of states like North Korea and Iran raise still more questions about nuclear security.

Not all uses of nuclear energy are so terrifying or complicated.[2] The power of the atom runs throughout our everyday lives: X-rays at routine visits to the dentist, americium-241 in smoke detectors, tritium in phosphorescent paints, and countless other isotopes put to use in the practice of science, medicine, and industry. Few people express concern about these technologies. Because most pay scant attention to the everyday uses of nuclear energy, its ubiquity does little to temper the fears and perceptions of risk raised by such topics as nuclear weapons, reactor meltdowns, and climate change.

These uncertain responses to nuclear energy—for electricity generation, as a weapon, and for other medical and industrial purposes—underscore a fundamental facet of the history of nuclear energy in the United States: Americans have never quite figured out what they think of it. Nuclear energy has evoked fears of mutant dystopias and hopes for brilliant technological futures; it has created both regimes of power and categories of victimization; it has been celebrated as a marker of humankind's mastery over nature just as it has been mourned as a sign of our ability to destroy the very fabric of life on earth. Public and academic understanding of nuclear energy has been both dualistic and paradoxical—that is, it seems to be two things at once, and contradictory things at that.

These conflicting and competing responses—what some scholars have called the "nuclear paradox"—have defined American responses to the technology since the dawn of the nuclear age. Americans displayed a full range of paradoxical responses to the news of the dropping of the first atomic bomb on Japan in August 1945. *New York Times* correspondent Anne O'Hare McCormick described "an explosion in men's minds as shattering as the obliteration of Hiroshima." Another commentator worried that "[for] all we know, we have created a Frankenstein," while the editors of *Life* magazine suggested that "Prometheus, the subtle artificer and friend of man, is still an American citizen."

Americans recognized a new era in the nation's—and the world's—environmental history, one in which humans had found a key to the laws of nature and unlocked an elemental power of huge magnitude. Seventy years later, even as climate change and the War on Terror have reframed nuclear choices, American reactions to nuclear energy are every bit as ambiguous as they were at the dawn of the nuclear age or at the height of the Cold War. Modern challenges have given nuclear issues a new sense of urgency, but the nuclear paradox remains unresolved. We—citizens, scientists, and academics alike—need to understand this paradox now more than ever.[3]

Why have American responses to nuclear energy been so complex and ambiguous? The documents collected in this volume suggest some answers to this basic question, but they also raise a host of other questions and avenues of inquiry about American society and its relationship to nature. Americans have struggled to define their responses to nuclear energy because of the way that it intersects with three of the most vital and contentious elements of postwar American life: changing relationships with nature, questions about the shape of civil society, and debates about the meaning of progress. These three themes run throughout the documents included in this book. The documents are not comprehensive or complete—they do not cover every aspect of nuclear energy, nor do they represent every important event or perspective. Rather, they focus on these three themes to help explain why responses to nuclear energy have been so contested and why modern solutions to nuclear dilemmas remain so elusive.

Analyzing primary documents requires paying attention to historical context—to the events and issues that informed and prompted the authors. This is a particularly challenging task for the documents included in this collection, because of the way that American ideas about nuclear energy changed over time and because of the way these encounters touched on so many different aspects of American life. To understand these documents, it is important to begin with an overview of the history of U.S. encounters with nuclear energy and a consideration of the way that this history shaped American thinking about nature, civil society, and the meaning of progress.

Nuclear energy played a central but underappreciated role in postwar American life. A "nuclear consensus" emerged in the early 1950s that bound together U.S. commitments to consumer society and nuclear energy. This consensus structured early American responses to the power of the atom, and it wove nuclear energy into the fabric of American life.

Historians have long recognized the key role of consumerism in postwar America. The historian Lizabeth Cohen has provided one of the most persuasive interpretations of this development, describing the emergence of what she labels a "Consumers' Republic"—that is, "an economy, culture, and politics built around the promises of mass consumption, both in terms of material life and the more idealistic goals of greater freedom, democracy, and equality." Mass consumption defined not just what one could purchase but also the ways in which one engaged in civil society. Through buying consumer goods, Americans received the benefits of washing machines and televisions and fulfilled their civic responsibilities. The Consumers' Republic became a strategy "for reconstructing the nation's economy and reaffirming its democratic values through promoting the expansion of mass consumption." The relationship between consumerism and citizenship informed American responses to the Cold War as well, as U.S. kitchens and garages filled with the latest model dishwashers and automobiles provided a powerful contrast to the material austerity of the Soviet Union. The Consumers' Republic lifted the American standard of living but also generated massive environmental consequences: mountains of garbage; urban, suburban, and wild landscapes threatened by pollution; and a consumer culture with a voracious appetite for natural resources extracted at home and abroad. Environmental concerns prompted some of the most significant critiques of the Consumers' Republic and its promises of progress through affluence.[4]

A commitment to nuclear energy lay at the center of the Consumers' Republic. Throughout the 1950s most Americans endorsed the emerging nuclear consensus: a diplomatic and military strategy based on nuclear weapons and the containment of communism and a vision of economic growth that welcomed nuclear energy both for the generation of electricity and for other peaceful and industrial uses. A nuclear shield protected consumer society and anchored a foreign policy that sought to open new markets for American consumer goods and to secure the natural resources necessary to fuel consumer society. Commercial nuclear power promised the cheap energy that seemed to put affluence within the reach of all Americans. Research and development spurred by advances in nuclear science produced many of the amenities that came to define the good life. This nuclear consensus intersected with other aspects of Cold War U.S. society and culture, such as beliefs about economic freedom, mass consumption, anticommunism, gender roles, and dissent.

The nuclear consensus also brought a host of environmental consequences. A foreign policy based on nuclear deterrence forced Americans to grapple with life in a world never far from war and to envision an apocalyptic future scarred by radioactivity and environmental catastrophe. Nuclear technologies for both military and domestic uses produced radioactive pollution—in the form of spent fuel, mill tailings, and fallout—that exacerbated the sense of environmental peril. Encounters with nuclear energy shaped the way that Americans thought about the natural world and provoked the modern environmental movement. Environmental risks, in turn, shaped the way that Americans responded to the nuclear consensus. By the late 1950s, a growing unease about the environmental consequences of nuclear energy provided a platform from which to challenge the consensus and all that it represented. The documents in this volume trace the building of the nuclear consensus and the challenge to it. In so doing, they reveal the central role of both nuclear energy and the environment in postwar U.S. history.

Several examples demonstrate the ways that the mutually reinforcing messages of the nuclear consensus and consumer society answered some of the early questions raised by nuclear energy, even as they raised new questions about the shape of civil society, the meaning of progress, and human relationships with nature. For example, the Soviet Union detonated its first atomic weapon in 1949, ending the brief period of American nuclear monopoly. President Harry S. Truman asked his National Security Council to consider the implications of this event for U.S. economic, diplomatic, and military interests. The council's response, known as NSC-68, neatly tied together the need for a permanent military, a large nuclear arsenal, a policy of containment in response to communism, and American economic expansion. NSC-68 declared the "fundamental purpose" of the United States to "assure the integrity and vitality of our free society, which is founded upon the dignity and worth of the individual." The Soviet Union stood in stark opposition to these U.S. goals, and assuring this fundamental purpose required "the free world to develop a successfully functioning political and economic system." It also required a powerful nuclear shield to protect that system. NSC-68 became one of the building blocks of American strategy during the Cold War and a key justification for U.S. participation in the arms race. In tying together American economic goals, nuclear strategy, and a particular vision of civil society, NSC-68 mirrored the goals of the Consumers' Republic, albeit with a military rather than an economic focus.[5]

The Atoms for Peace campaign, introduced by President Dwight Eisenhower in a 1953 speech before the General Assembly of the United Nations, underscored the paired messages of consumer society and the nuclear consensus. Eisenhower made clear the Cold War context of the campaign. The United States "wants itself to live in freedom, and in the confidence that the people of every other nation enjoy equally the right of choosing their own way of life." At the same time, Eisenhower sought to ease the tensions of the arms race. This meant a program dedicated to the development of commercial nuclear power, atomic energy for transportation and scientific research, and the use of radioisotopes in medicine, health, and agriculture. Low-cost nuclear power would strengthen the U.S. economy and further enable American-style economic development in other parts of the world—especially those areas targeted in the global battle against communism. The Atoms for Peace campaign received broad and deep support—from both the general public and from leaders of American business, labor, and government—all of whom recognized the importance of atomic energy for meeting the nation's domestic and foreign policy goals.[6]

The civil defense program of the 1950s reveals the intertwined natures of consumer society and the nuclear consensus through conceptions of civic responsibility, domesticity, and consumerism. As the arms race accelerated, the federal government urged citizens to prepare for a nuclear attack by building bomb shelters, joining home defense leagues, and learning skills such as firefighting and first aid. A significant portion of this campaign focused on women, reinforcing traditional, domestic gender roles as a way to provide stability and security in the face of the global uncertainties of the Cold War. Women could learn to bomb-proof their homes and practice "fire-safe housekeeping." They could also bolster the economy through their purchase of domestic consumer goods such as washing machines and dishwashers. In so doing, the campaign suggested, women assisted the nation in its struggle against communism. The civil defense program sent powerful cultural images of conservative gender roles and a consumer economy united in the global fight against communism, a unity that was both protected and threatened by nuclear Armageddon.[7]

The Atoms for Peace and civil defense programs raised questions about the shape of civil society—the rights and responsibilities of citizenship, the role of the government in promoting industry and protecting the environment, and the process of decision-making. These issues run throughout American encounters with nuclear energy. From the earliest days of the Manhattan Project, scientists, politicians, and soldiers wrestled with questions of how to make decisions about the use of nuclear weapons and the conduct of research on the uses of nuclear energy. Government bureaucrats, labor leaders, and industry executives debated the proper balance between government and industry in the development of nuclear power. Antinuclear activists worried that nuclear energy required a centralized, bureaucratic system of governance that itself represented a threat to democracy, and they also had very different ideas about the responsibilities of citizenship. These debates about the shape of civil society serve as an important theme of the documents in this book.

Both the Atoms for Peace program and domestic consumer culture rested on a linear view of progress and the assumption that scientific research and technological development would inevitably improve American material well-being. Atomic energy provided doctors with new tools for research, treatment, and diagnosis of cancer and a range of other medical conditions. Using radiation to spur mutations in crop plants would lead to more productive agriculture and food that would last longer. "Industrial radiation chemistry" promised harder and more durable plastics, longer-lasting automobile tires, and a host of other improved consumer goods. The power of the consumer economy and the advancements of science promised a higher standard of living for all. These promises lay at the heart of the nuclear consensus, predicated on the assumption that nuclear science pointed unerringly toward progress.

Nuclear energy provides a window into debates about the definition of progress, the value of affluence, and the role of science and technology in reaching these goals. These debates constitute a second theme of the documents, which offer a wide range of commentaries on the wisdom and folly of nuclear energy and of technological progress more generally. Some commentators saw in the atom a world of cheap energy that could free humans from manual labor and bring affluence to all; others saw a society dominated by a faithless technocratic elite, or nothing but environmental ruin. The nuclear consensus emerged at a time when most Americans trusted heavily in scientific expertise and other forms of authority. As American experience with nuclear energy deepened, this faith eroded. Debates about progress lay at the heart of the nuclear paradox: just what kind of future did this technology promise?

Nuclear energy brought radical changes to human relationships with nature, and these shifts serve as the third key theme of this book. "The age of ecology opened on the New Mexican desert, near the town of Alamogordo, on July 16, 1945, with a dazzling fireball of light and a swelling mushroom cloud of radioactive gasses," explains the environmental historian Donald Worster. Scholars in many disciplines saw in nuclear energy an opportunity to unlock the most basic secrets of nature. In seeking to understand patterns of fallout from weapons testing, for example, scientists developed a much more complete understanding of how energy flows through ecosystems by tracing radioactive isotopes. These insights, in turn, generated more questions and calls for new research. Howard and Eugene Odum, intellectual leaders of the developing field of ecosystem ecology and authors of the field's first textbook in 1953, developed their interest in ecology from their study of isotopes and eventually worked for the Atomic Energy Commission (AEC) analyzing the environmental impacts of weapons testing in the South Pacific.

Indeed, the AEC—the civilian governmental agency that oversaw the nation's nuclear program after 1946—spent huge sums of money for ecological research, helping to establish ecology as a major scientific field. Physics, oceanography, chemistry, and many other disciplines experienced similar revolutions as the result of scientific inquiry into questions posed by nuclear energy and warfare. In many cases, the U.S. Armed Forces provided funding for research that pushed these fields forward as military leaders sought any possible advantage over their Cold War adversaries. This meant advancements in mapping and modeling techniques and also research into the use of nature as a weapon—on how to provoke insect infestations and disease, devegetate vast areas, or alter the weather for strategic gain. The collaboration between the military and academic scientists generated a worldview that focused on environmental change and vulnerability. This worldview shaped the emergence of the modern environmental movement.[8]

Fallout—the radioactive dust created by weapons testing and broadcast into the atmosphere—prompted initial concerns about the environmental impacts of nuclear energy. Between 1951 and 1958 the United States detonated one hundred aboveground nuclear bombs at the Nevada Test Site in southern Nevada and hundreds more in subsequent decades either underground in Nevada or on remote islands in the Pacific Ocean. Although the environmental implications of fallout were more intense closer to the test sites, fallout drifted eastward and became a national problem. The presence of the radioactive isotope Strontium-90 in milk emerged as one of the most mobilizing aspects of the issue. Strontium mimics calcium in the way it concentrates in bones, and in the mid-1950s the general public became increasingly alarmed about the safety of milk and other dairy products. The "Baby Tooth Survey" initiated by the biologist Barry Commoner and the Greater St. Louis Citizens' Committee for Nuclear Information in 1958 fused research and education in a way that crystalized public concern about fallout. The study's authors encouraged children to send their baby teeth in for testing on the presence of radioactive isotopes. Results of the survey revealed that Strontium-90 levels in the teeth of children born in the 1950s had risen steadily during the decade. No longer confined to Japanese cities or Nevada deserts, fallout had invaded the kitchens and nurseries of everyday America.[9]

Concerns about fallout both foreshadowed and galvanized the modern environmental movement. Historians frequently date the origin of the environmental movement to the publication of Rachel Carson's 1962 classic, *Silent Spring*. The book raised awareness about chemical pesticides, introduced the public to the emerging discipline of ecology, and caused people around the nation and the world to question the environmental costs of progress. The book was also laced with concerns about fallout and radioactivity. It opened with the "Fable for Tomorrow," in which a "fine white dust" causes a blight to descend on an idyllic American town. The first threats that Carson mentioned specifically are nuclear, not chemical. "Strontium 90, released through nuclear explosions into the air," she wrote, "comes to earth in rain or drifts down as fallout, lodges in soil, enters into the grass or corn or wheat grown there, and in time takes up its abode in the bones of a human being, there to remain until his death." The insidious danger Carson depicted throughout *Silent Spring*—of invisible poisons moving through the food chain and causing unknown and unanticipated damage to human health and the environment—applied equally well to radioactive isotopes as to industrial pesticides. Some scholars have attributed the book's remarkable popularity to a public primed for its message by fears of fallout. The recognition of the human capability to cause catastrophic environmental damage, and even human extinction, became a frequent refrain in the environmental and antinuclear movements. So, too, did Carson's questioning of the mixed blessings of the "progress" promised by consumer society and the nuclear consensus.[10]

Yet not all environmentalists opposed nuclear energy, and the environmental movement displayed a range of reactions to the technology. In addition to prompting developments in ecological science and raising the specter of human extinction, nuclear energy offered other paths to both ecological salvation and ruin. Some environmentalists embraced nuclear energy as a cleaner alternative to the fossil fuels that caused urban air pollution, acid rain, and climate change, while others bitterly opposed commercial nuclear energy for its creation of long-lasting radioactive waste and the way that it fostered a powerful, centralized technological bureaucracy. Questions about how to respond to the promise and peril of nuclear energy often divided environmental groups. For example, David Brower was forced out of his role as executive director of the Sierra Club for his antinuclear positions. From its earliest years through more recent debates over climate change, nuclear energy has split the environmental community. That even those with similar goals and perspectives disagree over the potential, and potential consequences, of this technology speaks to the power of the nuclear paradox.[11]

The environmental and human health questions raised by fallout, the arms race, and commercial nuclear power had another crucial consequence: they provided a platform for the first challenges to the nuclear consensus. At first, only a few people questioned the consensus—and those who did suffered harsh consequences. J. Robert Oppenheimer, for example, challenged the decision to develop the hydrogen bomb and subsequently had his security clearance revoked after highly publicized hearings before the House Un-American Activities Committee raised questions about his past affiliation with the Communist Party and his current loyalties. In 1955 eleven prominent scientists and intellectuals, led by the physicist Albert Einstein and the philosopher Bertrand Russell, issued a much more sustained and direct challenge to what they perceived as the illogic of the accelerating arms race. Other, more grassroots peace movements followed as a cultural space that allowed a challenge to the consensus opened. This plea had an essential environmental theme: to view humans as a biological species united by the threats posed by nuclear energy, not as separate peoples divided by ideology. The international peace, environmental, and antinuclear movements of the next several decades frequently echoed this logic.[12]

By the early 1960s these concerns had grown more commonplace as the challenge to the nuclear consensus gained steam. Two events during the presidency of John F. Kennedy raised further questions about the nuclear consensus: the Cuban Missile Crisis in 1962 and the signing of the Partial Test Ban Treaty in 1963. The missile crisis heightened fears of a nuclear conflict and prompted many Americans to rethink some of the basic assumptions of the Cold War and the arms race. It also prompted President Kennedy to call for improved relations with the Soviet Union. His position rested in part on environmental concerns. "Total war makes no sense . . . in an age when the deadly poisons produced by a nuclear exchange would be carried by wind and water and soil and seed to the far corners of the globe and to generations yet unborn." The Test Ban Treaty followed soon after, banning atmospheric nuclear weapons testing—one of the first tangible steps toward arms control.[13]

The missile crisis and the test ban prompted a sharp and unexpected turn in American thinking about nuclear weapons and the arms race. After nearly fifteen years of increasing anxieties about fallout and nuclear war, these fears diminished. Manhattan Project scientist Eugene Rabinowitch noticed: "A year ago, we had just escaped nuclear war over the Soviet missiles in Cuba. People saw in a flash how close we are living to the edge of the precipice. Peace movements flourished and disarmament studies proliferated. . . . This acute concern and frantic search for solutions did not last long." Historians have suggested several reasons for the decline in public concern about nuclear technologies in the 1960s, including the perception of reduced danger, the success of the Atoms for Peace program, and a shift in focus to other issues such as the civil rights movement and the Vietnam War. Although nuclear fear might have lessened, the arms race hurtled on. The design, construction, and deployment of weapons with ever-greater destructive power and accuracy continued.[14]

At the same time, the commitment of government and industry to nuclear power spiked as the first reactors came online. Commercial nuclear power started slowly, despite the high hopes of industry boosters. By the late 1960s utilities were finally willing to invest in nuclear technologies. This economic calculus came in response to the environmental movement's concerns about air quality and an expected rise in the price of fossil fuels due to the cost of pollution control. Industry experts predicted that American demand for electricity would continue to increase—a concrete tie to the still expanding consumer economy. The energy crisis of the mid-1970s, spurred by the 1973 embargo on exporting oil to the United States by the Organization of Petroleum Exporting Countries, further heightened calls for nuclear power as a way to bolster American energy independence. In the period between 1955 and 1965, utilities ordered 22 nuclear units. This number surged to 68 between 1966 and 1968 and to 106 between 1971 and 1973. The AEC predicted that the nation would have a thousand nuclear-generating plants by the year 2000. Industry watchers called it a "bandwagon market."[15]

The expansion of commercial nuclear power forced Americans to confront different aspects of the nuclear paradox and to balance more directly the promises and risks of industrial uses of nuclear energy. Beginning in the late 1960s and accelerating in the 1970s, an antinuclear movement coalesced in opposition to the predicted surge in reactor development. The motivations for this opposition were complex—antinuclear protesters drew on environmental issues but supplemented these with other concerns about centralized authority and participatory democracy. The historian Thomas Wellock characterized the antinuclear movement as "a broad coalition opposed to the political authority behind the peaceful atom. . . . [offering] a fundamental critique of government, industry, and scientific authority. The public, they believed, had been seduced into giving up their rights by a 'small elite corps of nuclear experts' and persuaded to an 'abject worship of technology' through the seductive language of progress. . . . [Nuclear] power served as a symbol of the moral and political failure of the modern American state." In short, this coalition opposed the vision of civil society and progress advanced by the nuclear consensus. Antinuclear groups conducted high-profile direct action protests at Prairie Island in Minnesota, Seabrook Station in New Hampshire, and Diablo Canyon in California. While unsuccessful in stopping these particular plants, the antinuclear movement—along with industrial mismanagement, technical failures, rising construction costs, the intractable problem of nuclear waste, and changing economic forecasts—contributed to a drastic reversal of fortune for the industry. By the time that the partial meltdown at Three Mile Island in 1979 evaporated public support for nuclear energy, orders for new reactors had ceased and the nuclear industry had come to a grinding a halt. All of these factors contributed to the industry's problems, but Three Mile Island was a pivotal event in shaping public perception about commercial nuclear power.[16]

The ebb in nuclear fear that followed the Test Ban Treaty in 1963 abruptly ended with the election of Ronald Reagan as president in 1980. Reagan sought to renew the nuclear consensus—once again linking economic and political freedoms both at home and abroad to a strong economy and an expanded military with an enhanced nuclear arsenal. Reagan identified two threats to freedom and progress: an overreaching government that stifled American economic development and a totalitarian regime in the Soviet Union. Nuclear energy could counter both threats. In response to perceived Russian military superiority—and what he famously labeled "the aggressive impulses of an evil empire"—Reagan initiated the most significant peacetime buildup of military capacity in American history. The path to peace, he argued, lay through military strength. At the same time, Reagan called for a renewed commitment to commercial nuclear power, a commitment that he tied to his goals of economic deregulation and federal downsizing. He declared a "more abundant, affordable, and secure energy future for all Americans" as a key element of his administration's economic goals. Nuclear energy could meet this need, Reagan explained, but only if the government allowed private industry to operate freely. "Unfortunately, the Federal Government has created a regulatory environment that is forcing many utilities to rule out nuclear power as a source of new generating capacity. . . . Nuclear power has become entangled in a morass of regulations that do not enhance safety but that do cause extensive licensing delays and economic uncertainty." Reagan proved more effective in reinstating military superiority than resuscitating the nuclear power industry; stealth bombers, Trident submarines, and tactical weapons enhanced the nation's nuclear arsenal but commercial nuclear power remained moribund, beset by high operating costs and public opposition. The Reagan presidency brought more than just rhetoric, and tensions between the superpowers escalated to levels not seen since the Cuban Missile Crisis.[17]

These tensions prompted a resurgence of nuclear fear as well as a renewed peace movement. A postnuclear world of environmental catastrophe once again seemed easy to envision. The fear of nuclear war reemerged on the cultural scene, with a bevy of novels, movies, and cultural commentaries that imagined the environmental and human consequences of a nuclear war that Reagan's aggressive rhetoric seemed to make more likely. Scientists used computer models to predict "nuclear winter"—the debris injected into the atmosphere by multiple nuclear explosions would cool temperatures, end agriculture, and make the survivors of a confrontation envy the dead. These fears sparked the Nuclear Freeze campaign, an effort to get both superpowers to keep the number of weapons at their current levels. The movement enjoyed broad public support and sparked the largest political rally in U.S. history when one million people converged on Central Park in New York City in 1982. Several states and many municipalities passed nuclear freeze resolutions, as did the U.S. House of Representatives, although the measure died in the Senate.[18]

The increasing tensions and rhetoric had a resolution that seemed unimaginable at the start of the Reagan presidency: the end of the Cold War. Historians have debated the role of the Reagan administration's policies in this development, but in the mid-1980s the United States and the U.S.S.R. negotiated on a range of arms control topics. With both countries responding to domestic pressures to reach an agreement, in 1987 negotiators agreed on the Intermediate-Range Nuclear Forces Treaty, eliminating intermediate range, missile-launched weapons and removing many of these weapons from Europe. These talks coincided with radical social and economic changes in the Soviet Union—and its eventual dissolution. The opening of the Berlin Wall in 1989 symbolized the end of the Cold War—and the end of an era in American encounters with nuclear energy.

THE NUCLEAR PAST AND THE NUCLEAR FUTURE

While the end of the Cold War stopped the arms race, it also inaugurated a period of what might best be described as nuclear confusion. Since the 1950s, the idea of deterrence had helped to keep the peace—if at least by making the concept of nuclear war unthinkable. But in a post–Cold War world—a world without two equally balanced powers—what would maintain this equilibrium? Nuclear strategists pondered how to defend against new kinds of threats, those posed by terrorism and rogue states. Old issues like nuclear waste remained unresolved—and aging stockpiles of nuclear weapons contributed to the problem. At the same time, climate change emerged as the single greatest environmental threat, and many environmentalists began to give nuclear energy a fresh look as a post-carbon-energy source.

The patterns of thought defined by the rise and fall of the nuclear consensus no longer shape encounters with nuclear energy, but the nuclear paradox remains. The same issues that have always made American responses to nuclear energy so complicated—questions about human relations with nature, the shape of civil society, and the meaning of progress—lie at the center of modern debates about nuclear energy and climate change. The politicization of science and questions about how to apply scientific conclusions to public policy cloud responses to climate change, much as they did to Americans' understanding of fallout and the risks of nuclear power. Climate change is so contentious because it seems to pit the needs of human development against environmental health. How can we both stabilize the climate and guarantee a reasonable standard of living—not just in the United States but also in the developing world? Should the goal of energy and development policies be the preservation of American affluence and its extension around the globe? Some people see this type of development as progress and view nuclear energy as a necessary driver of low-carbon economic growth. Others fear that such development would bring an environmental catastrophe on par with Cold War predictions of nuclear disaster and suggest that climate mitigation demands a retreat from the promise of affluence that has served as the bedrock of American consumer society. These questions are not new; balancing economic development, environmental risk, and the promise of affluence has been a central dilemma of American encounters with nuclear energy.

Anxieties about nuclear apocalypse have calmed, and the presence of nuclear energy in our lives is harder to see and harder to recognize. But the nuclear paradox remains woven into the fabric of our daily lives and conditions the decisions that we face today about how to use (or not use) nuclear energy. Nuclear fuels powered the dishwashers and televisions that came to define affluence in postwar America; most of those power plants continue to run. A nuclear shield enabled and protected the development of that affluence—and it continues to do so. Nuclear energy shaped the roads we drive on, the goods we purchase, and the health of our environments. The spent fuel that continues to build up at reactors around the country serves as a powerful symbol of the environmental trade-offs embedded within consumer society—trade-offs that include both convenience and waste, affluence and inequality, cheap energy and a changing climate.

The documents in this collection connect the nuclear past to the nuclear future. They provide evidence of how people have debated the thorny issues raised by nuclear energy; many of these debates have resurfaced as Americans take a new look at nuclear technology. As you read the documents, ask of them the same questions required in all historical analysis: Who wrote these documents, and why? Who was the intended audience, and how might this have shaped the authors' messages and goals? Pay particular attention to the assumptions that authors made about relationships with nature, the shape of civil society, and the definition of progress. How did different stakeholders invoke such terms as *nature*, *citizenship*, and *progress*, and to what end? What patterns connect assumptions, stakeholders, and policy positions? On the one hand, we need to be careful about relating conclusions drawn from the past to the modern situation. The end of the Cold War, the War on Terror, new prospects for renewable energy, and climate change have created a very different set of conditions that have generated different responses to nuclear energy. On the other hand, we wrestle with many of the same issues as we did at the dawn of the Atomic Age. Do the patterns of the past apply to modern attempts to resolve the nuclear paradox? How have American ideas about nature, civil society, and progress changed over time? The nuclear past offers clues to how we have discussed these issues before, and how we might do so again.

NOTES

1. Ann S. Bisconti, "Upward Trend in Public's Favorable Attitude toward Nuclear Energy," *Perspectives in Public Opinion* (April 2013), available at www.nei.org/CorporateSite/media/filefolder/NEI-Perspective-On-Public-Opinion_April-2013.pdf?ext=.pdf; Amelia Urry, "Don't Panic! Fukushima Radiation Just Hit the West Coast," *Grist*, January 7, 2015, available at http://grist.org/news/dont-panic-fukushima-radiation-just-hit-the-west-coast/; and Yale Project on Climate Change Communication, "Nuclear Power in the American Mind," available at http://environment.yale.edu/climate-communication/article/nuclear-power-in-the-american-mind, 2012.
2. Throughout this book I use the term *nuclear power* to refer to the generation of electricity from nuclear reactions and the term *nuclear energy* to refer to all uses of the atom—as a weapon, to generate electricity, and for other medical, scientific, and industrial uses.
3. Spencer R. Weart, *The Rise of Nuclear Fear* (Cambridge, MA: Harvard University Press, 2012); and Paul Boyer, *By the Bomb's Early Light: American Thought and Culture at the Dawn of the Atomic Age* (New York: Pantheon Books, 1985), xix, 4–9.
4. Lizabeth Cohen, *A Consumers' Republic: The Politics of Mass Consumption in Postwar America* (New York: Vintage Books, 2003), 7, 11.
5. "NSC 68: United States Objectives and Programs for National Security," *Naval War College Review* 28 (May–June 1975): 51–108, quotations on 54 and 98.
6. Dwight D. Eisenhower, "Atoms for Peace," December 8, 1953, available from the Dwight D. Eisenhower Presidential Library, Museum, and Boyhood Home at www.eisenhower.archives.gov/research/online_documents/atoms_for_peace/Binder13.pdf; and Richard G. Hewlett and Jack M. Holl, *Atoms*

for Peace and War, 1953–1961: Eisenhower and the Atomic Energy Commission (Berkeley: University of California Press, 1989), 209–10, 240.

7 Elaine Tyler May, *Homeward Bound: American Families and the Cold War* (New York: Basic Books, 1988), 159, 167–68.

8 Donald Worster, *Nature's Economy: A History of Ecological Ideas* (New York: Cambridge University Press, 1977), 362–63; Laura A. Bruno, "The Bequest of the Nuclear Battlefield: Science, Nature, and the Atom during the First Decade of the Cold War," *Historical Studies in the Physical and Natural Sciences* 33, no. 2 (2003): 237–60; and Jacob Darwin Hamblin, *Arming Mother Nature: The Birth of Catastrophic Environmentalism* (New York: Oxford University Press, 2013), 11.

9 Michael Egan, *Barry Commoner and the Science of Survival: The Remaking of American Environmentalism* (Cambridge, MA: MIT Press, 2007).

10 Rachel Carson, *Silent Spring* (Boston: Houghton Mifflin Company, 1962), 2–3, 6; and Ralph H. Lutts, "Rachel Carson's *Silent Spring*, Radioactive Fallout, and the Environmental Movement," *Environmental Review* 9 (Autumn 1985): 210–25.

11 Thomas Raymond Wellock, *Critical Masses: Opposition to Nuclear Power in California, 1958–1978* (Madison: University of Wisconsin Press, 1998).

12 Kai Bird and Martin J. Sherwin, *American Prometheus: The Triumph and Tragedy of J. Robert Oppenheimer* (New York: Alfred A. Knopf, 2005); and Bertrand Russell and Albert Einstein, "The Russell-Einstein Manifesto," July 9, 1955, Pugwash Conferences on Science and World Affairs, available at http://pugwash.org/1955/07/09/statement-manifesto/.

13 John F. Kennedy, "Commencement Address at American University in Washington," June 10, 1963, available at www.presidency.ucsb.edu/ws/?pid=9266.

14 Eugene Rabinowitch, "New Year's Thoughts 1964," *Bulletin of the Atomic Scientists* 20 (January 1964): 2; and Paul S. Boyer, *Fallout: A Historian Reflects on America's Half-Century Encounter with Nuclear Weapons* (Columbus: Ohio State University Press, 1998), 111–23.

15 J. Samuel Walker, *Three Mile Island: A Nuclear Crisis in Historical Perspective* (Berkeley: University of California Press, 2004), 6.

16 Wellock, *Critical Masses*, 9.

17 Ronald Reagan, "Remarks at the Annual Convention of the National Association of Evangelicals in Orlando, Florida," March 8, 1983, available at www.reagan.utexas.edu/archives/speeches/1983/30883b.htm, and "Statement Announcing a Series of Policy Initiatives on Nuclear Energy," October 8, 1981, available at www.presidency.ucsb.edu/ws/?pid=44353.

18 Weart, *Rise of Nuclear Fear*, 294–95.

PART I

FIRST REACTIONS

Americans first encountered nuclear energy a generation before World War II. European scientists such as Wilhelm Conrad Roentgen and Marie Curie had experimented with radioactivity and X-rays since the 1890s; Curie won two Nobel Prizes for her explorations of radiation. Radium-infused paints hit the U.S. market in the late 1910s and became key features on everything from watch dials to children's toys, their glow-in-the-dark fluorescence representing the arrival of modernity. That the young women who applied radium paint to watch dials eventually suffered from radiation poisoning also came to stand for modernity's mixed blessings. By the 1930s, radioactivity and atomic imagery claimed a central role in American thinking about science and technology, modernity and progress.

The August 1945 bombing of Hiroshima and Nagasaki, however, represented something different. The documents in part 1 demonstrate the range of American reactions to the newly arrived Atomic Age. The nuclear consensus had not yet coalesced; there was no clear path for answering the questions raised by nuclear energy. What kinds of issues were Americans grappling with? What were they worried about? How did people fit nuclear energy into their ideas about nature, civil society, and progress? In what ways do these first reactions to nuclear energy foreshadow the nuclear consensus that emerged in the 1950s—and which point to paths not taken?

LESLIE GROVES

REPORT ON THE TRINITY TEST, 1945

Lieutenant General Leslie Groves served as the military commander who oversaw the Manhattan Project—the top-secret wartime project that developed atomic bombs. In the document below, Groves reports to his military superiors about the Trinity Test—the first successful detonation of a nuclear bomb. His report also includes the observations of General Thomas F. Farrell. How do the two military men describe the blast? What comparisons do they use, what language and images do they convey? How do these two accounts differ? What did they see as the long-term significance of the atomic bomb?

2. At 0530, 16 July 1945, in a remote section of the Alamogordo Air Base, New Mexico, the first full scale test was made of the implosion type atomic fission bomb. For the first time in history there was a nuclear explosion. And what an explosion. . . . The bomb was not dropped from an airplane but was exploded on a platform on top of a 100-foot high steel tower.

3. The test was successful beyond the most optimistic expectations of anyone. Based on the data which it has been possible to work up to date, I estimate the energy generated to be in excess of the equivalent of 15,000 to 20,000 tons of TNT; and this is a conservative estimate. Data

Excerpted from Leslie Groves, "Memorandum for the Secretary of War," *Foreign Relations of the United States: The Conference of Berlin (The Potsdam Conference), 1945*, vol. 2 (Washington, D.C.: Government Printing Office, 1960), 1361–68.

based on measurements which we have not yet been able to reconcile would make the energy release several times the conservative figure. There were tremendous blast effects. For a brief period there was a lighting effect within a radius of 20 miles equal to several suns in midday; a huge ball of fire was formed which lasted for several seconds. This ball mushroomed and rose to a height of over ten thousand feet before it dimmed. The light from the explosion was seen clearly at Albuquerque, Santa Fe, Silver City, El Paso and other points generally to about 180 miles away. The sound was heard to the same distance in a few instances but generally to about 100 miles. Only a few windows were broken although one was some 125 miles away. A massive cloud was formed which surged and billowed upward with tremendous power, reaching the substratosphere at an elevation of 41,000 feet, 36,000 feet above the ground in about five minutes, breaking without interruption through a temperature inversion at 17,000 feet which most of the scientists thought would stop it. Two supplementary explosions occurred in the cloud shortly after the main explosion. . . . Huge concentrations of highly radioactive materials resulted from the fission and were contained in this cloud. . . .

5. One-half mile from the explosion there was a massive steel test cylinder weighing 220 tons. The base of the cylinder was solidly encased in concrete. . . . The blast tore the tower from its foundations, twisted it, ripped it apart and left it flat on the ground. The effects on the tower indicate that, at that distance, unshielded permanent steel and masonry buildings would have been destroyed. I no longer consider the Pentagon a safe shelter from such a bomb. . . .

6. The cloud traveled to a great height first in the form of a ball, then mushroomed, then changed into a long trailing chimney-shaped column and finally was sent in several directions by the variable winds at the different elevations. It deposited its dust and radioactive materials over a wide area. It was followed and monitored by medical doctors and scientists with instruments to check its radioactive effects. While here and there the activity on the ground was fairly high, at no place did it reach a concentration which required evacuation of the population. Radioactive material in small quantities was located as much as 120 miles away. The measurements are being continued in order to have adequate data with which to protect the Government's interests in case

of future claims. For a few hours I was none too comfortable about the situation. . . .

11. Brigadier General Thomas F. Farrell was at the control shelter located 10,000 yards south of the point of explosion. His impressions are given below:

"The scene inside the shelter was dramatic beyond words. In and around the shelter were some twenty-odd people concerned with last minute arrangements prior to firing the shot. Included were: Dr. [J. Robert] Oppenheimer, the Director who had borne the great scientific burden of developing the weapon from the raw materials made in Tennessee and Washington and a dozen of his key assistants—Dr. [George] Kistiakowsky, who developed the highly special explosives; Dr. [Kenneth] Bainbridge, who supervised all the detailed arrangements for the test; Dr. [Jack M.] Hubbard, the weather expert, and several others. Besides these, there were a handful of soldiers, two or three Army officers and one Naval officer. The shelter was cluttered with a great variety of instruments and radios. . . .

"Just after General Groves left, announcements began to be broadcast of the interval remaining before the blast. They were sent by radio to the other groups participating in and observing the test. As the time interval grew smaller and changed from minutes to seconds, the tension increased by leaps and bounds. Everyone in that room knew the awful potentialities of the thing that they thought was about to happen. The scientists felt that their figuring must be right and that the bomb had to go off but there was in everyone's mind a strong measure of doubt. The feeling of many could be expressed by 'Lord, I believe; help Thou mine unbelief.' We were reaching into the unknown and we did not know what might come of it. It can be safely said that most of those present—Christian, Jew and Atheist—were praying and praying harder than they had ever prayed before. If the shot were successful, it was a justification of the several years of intensive effort of tens of thousands of people—statesmen, scientists, engineers, manufacturers, soldiers, and many others in every walk of life.

"In that brief instant in the remote New Mexico desert the tremendous effort of the brains and brawn of all these people came suddenly and startlingly to the fullest fruition. Dr. Oppenheimer, on whom had rested a very heavy burden, grew tenser as the last seconds ticked off.

He scarcely breathed. He held on to a post to steady himself. For the last few seconds, he stared directly ahead and then when the announcer shouted 'Now!' and there came this tremendous burst of light followed shortly thereafter by the deep growling roar of the explosion, his face relaxed into an expression of tremendous relief. Several of the observers standing back of the shelter to watch the lighting effects were knocked flat by the blast.

"The tension in the room let up and all started congratulating each other. Everyone sensed 'This is it!' No matter what might happen now all knew that the impossible scientific job had been done. Atomic fission would no longer be hidden in the cloisters of the theoretical physicists' dreams. It was almost full grown at birth. It was a great new force to be used for good or for evil. There was a feeling in that shelter that those concerned with its nativity should dedicate their lives to the mission that it would always be used for good and never for evil.

"Dr. Kistiakowsky, the impulsive Russian, threw his arms around Dr. Oppenheimer and embraced him with shouts of glee. Others were equally enthusiastic. All the pent-up emotions were released in those few minutes and all seemed to sense immediately that the explosion had far exceeded the most optimistic expectations and wildest hopes of the scientists. All seemed to feel that they had been present at the birth of a new age—The Age of Atomic Energy—and felt their profound responsibility to help in guiding into right channels the tremendous forces which had been unlocked for the first time in history.

"As to the present war, there was a feeling that no matter what else might happen, we now had the means to insure its speedy conclusion and save thousands of American lives. As to the future, there had been brought into being something big and something new that would prove to be immeasurably more important than the discovery of electricity or any of the other great discoveries which have so affected our existence.

"The effects could well be called unprecedented, magnificent, beautiful, stupendous and terrifying. No man-made phenomenon of such tremendous power had ever occurred before. The lighting effects beggared description. The whole country was lighted by a searing light with the intensity many times that of the midday sun. It was golden, purple, violet, gray and blue. It lighted every peak, crevasse and ridge of the nearby mountain range with a clarity and beauty that cannot be described but must be seen to be imagined. It was that beauty the great poets dream about but describe most poorly and inadequately. Thirty seconds after the explosion came first, the air blast pressing hard against the people and things, to be followed almost immediately by the strong, sustained, awesome roar which warned of doomsday and made us feel that we puny things were blasphemous to dare tamper with the forces heretofore reserved to The Almighty. Words are inadequate tools for the job of acquainting those not present with the physical, mental and psychological effects. It had to be witnessed to be realized."

12. My impressions of the night's high points follow:

. . . At about two minutes of the scheduled firing time all persons lay face down with their feet pointing towards the explosion. As the remaining time was called from the loud speaker from the 10,000 yard control station there was complete silence. . . . There was then this burst of light of a brilliance beyond any comparison. We all rolled over and looked through dark glasses at the ball of fire. About forty seconds later came the shock wave followed by the sound, neither of which seemed startling after our complete astonishment to the extraordinary lighting intensity. Dr. [James] Conant reached over and we shook hands in mutual congratulations. Dr. [Vannevar] Bush, who was on the other side of me, did likewise. The feeling of the entire assembly was similar to that described by General Farrell, with even the uninitiated feeling profound awe. Drs. Conant and Bush and myself were struck by an even stronger feeling that the faith of those who had been responsible for the initiation and the carrying on of this Herculean project had been justified. I personally thought of Blondin crossing Niagara Falls on his tight rope, only to me this tight rope had lasted for almost three years and of my repeated confident-appearing assurances that such a thing was possible and that we would do it.

HARRY S. TRUMAN

WHITE HOUSE STATEMENT ON THE BOMBING OF HIROSHIMA, 1945

On August 3, 1945, President Harry Truman authorized the use of the newly developed bombs on the Japanese mainland. Three days later, the U.S. Army Air Force dropped the first bomb on the Japanese city of Hiroshima. Most Americans learned of the existence of atomic bombs from a statement issued by President Truman shortly after the bombing. He broadcast the press release over the radio, and newspapers around the country republished or summarized the statement. How does Truman describe the bomb and its power—what terms and metaphors does he employ, and why did he use these terms? How might the average American have understood this new technology? What benefits and risks does Truman predict from the use of atomic weapons?

Sixteen hours ago an American airplane dropped one bomb on Hiroshima and destroyed its usefulness to the enemy. That bomb had more power than 20,000 tons of T.N.T. It had more than two thousand times

---

Excerpted from Harry S. Truman, White House Press Release on Hiroshima, August 6, 1945, Harry S. Truman Library and Museum, www.trumanlibrary.org/whistlestop/study_collections/bomb/large/documents/index.php?documentdate=1945-08-06&documentid=59&pagenumber=1.

the blast power of the British "Grand Slam" which is the largest bomb ever yet used in the history of warfare.

The Japanese began the war from the air at Pearl Harbor. They have been repaid many fold. And the end is not yet. With this bomb we have now added a new and revolutionary increase in destruction to supplement the growing power of our armed forces. In their present form these bombs are now in production and even more powerful forms are in development.

It is an atomic bomb. It is a harnessing of the basic power of the universe. The force from which the sun draws its power has been loosed against those who brought war to the Far East.

Before 1939, it was the accepted belief of scientists that it was theoretically possible to release atomic energy. But no one knew any practical method of doing it. By 1942, however, we knew that the Germans were working feverishly to find a way to add atomic energy to the other engines of war with which they hoped to enslave the world. But they failed. We may be grateful to Providence that the Germans got the V-1's and V-2's late and in limited quantities and even more grateful that they did not get the atomic bomb at all.

The battle of the laboratories held fateful risks for us as well as the battles of the air, land and sea, and we have now won the battle of the laboratories as we have won the other battles.

Beginning in 1940, before Pearl Harbor, scientific knowledge useful in war was pooled between the United States and Great Britain, and many priceless helps to our victories have come from that arrangement. Under that general policy the research on the atomic bomb was begun. With American and British scientists working together we entered the race of discovery against the Germans. . . . We have spent two billion dollars on the greatest scientific gamble in history—and won.

But the greatest marvel is not the size of the enterprise, its secrecy, nor its cost, but the achievement of scientific brains in putting together infinitely complex pieces of knowledge held by many men in different fields of science into a workable plan. And hardly less marvelous has been the capacity of industry to design, and of labor to operate, the machines and methods to do things never done before so that the brainchild of many minds came forth in physical shape and performed as it was supposed to do. Both science and industry worked under the direction of the United States Army, which achieved a unique success in managing so diverse a problem in the advancement of knowledge in an amazingly short time. It is doubtful if such another combination could be got together in the world. What has been done is the greatest achievement of organized science in history. It was done under high pressure and without failure.

We are now prepared to obliterate more rapidly and completely every productive enterprise the Japanese have above ground in any city. We shall destroy their docks, their factories, and their communications. Let there be no mistake; we shall completely destroy Japan's power to make war.

It was to spare the Japanese people from utter destruction that the ultimatum of July 26 was issued at Potsdam. Their leaders promptly rejected that ultimatum. If they do not now accept our terms they may expect a rain of ruin from the air, the like of which has never been seen on this earth. Behind this air attack will follow sea and land forces in such numbers and power as they have not yet seen and with the fighting skill of which they are already well aware. . . .

The fact that we can release atomic energy ushers in a new era in man's understanding of nature's forces. Atomic energy may in the future supplement the power that now comes from coal, oil, and falling water, but at present it cannot be produced on a basis to compete with them commercially. Before that comes there must be a long period of intensive research.

It has never been the habit of the scientists of this country or the policy of this Government to withhold from the world scientific knowledge. Normally, therefore, everything about the work with atomic energy would be made public.

But under the present circumstances it is not intended to divulge the technical processes of production or all the military applications, pending further examination of possible methods of protecting us and the rest of the world from the danger of sudden destruction.

I shall recommend that the Congress of the United States consider promptly the establishment of an appropriate commission to control the production and use of atomic power within the United States. I shall give further consideration and make further recommendations to the Congress as to how atomic power can become a powerful and forceful influence towards the maintenance of world peace.

NAGASAKI MUSHROOM CLOUD, 1945

Americans learned of the atomic bomb from President Truman's press release. But they first saw it in images like this one, published in the popular magazine *Life* and many other places in August 1945. The mushroom cloud quickly became the iconic image of the Atomic Age. How might people have reacted to this image, upon seeing it for the first time?

. . . .

U.S. Department of Energy, *Manhattan Project: An Interactive History*, www.osti.gov/opennet/manhattan-project-history/Resources/photo_gallery/nagasaki_images.htm.

JOSEPH H. WILLITS

"SOCIAL ADJUSTMENTS TO ATOMIC ENERGY," 1946

Economist Joseph H. Willits worked for the Rockefeller Foundation, a philanthropic organization dedicated to improving the human condition throughout the world. Willits worried that the nation's social and governmental institutions would be unable to cope with the challenge to civil society posed by nuclear technologies. What kind of institutions did Willits view as necessary to deal with atomic energy, and why did these institutions worry him? What did he see as the social and political implications of nuclear energy? Why does he regard it with such ambivalence? His mention of the May-Johnson Bill refers to an early proposal of the Atomic Energy Act, one that envisioned a much greater involvement by the military in the administration of the federal atomic energy program.

....

The history of inventions shows that at the time of a particular discovery the social effects to come from that invention can only be judged as the size of an iceberg is judged with seven-eighths of its whole below the surface of visibility. Natural scientists have as yet provided us very little of the essential concrete information on atomic energy; so a social scientist who comments on the social implications of atomic energy cannot speak about

---

Excerpted from Joseph H. Willits, "Social Adjustments to Atomic Energy," *Proceedings of the American Philosophical Society* (January 1, 1946): 48–52. Courtesy of the American Philosophical Society.

results but only about the process of social adjustment to the changes, present and prospective, set in motion by the release of that energy. . . .

INTERNATIONAL RELATIONS

The advance of science and technology—the atomic bomb is merely one episode in that advance—poses an old choice with a new and terrifying urgency. Modern society shall avoid war or war shall annihilate modern society. Atomic energy may enable man to destroy himself. The great hope is that man will perceive his danger and act while he still has the power to guard against catastrophe.

There is no way to guarantee peace. Even if a successful attempt were made to win world unification by conquest, such a world would not be peaceful, democratic, nor free. I see no hope of world government by formula. We have to work from where we are, not from where we wish we were. The slow building and strengthening of a world community that will settle its disputes without resort to war is a process that cannot be achieved by a simple tour de force. It requires the efforts of scholars and scientists, of experts and administrators, of statesmen and political leaders, of teachers, both popular and academic, each contributing in his appropriate way. But it is possible to go beyond such general assurance and, in international policy, to identify certain specific opportunities for emphasis which are of critical importance in efforts to build a constructive and durable peace. . . .

COMMERCIAL ATOMIC POWER

Since it is possible, by human intervention, to slow down the rate of atomic fission, I assume it to be only a matter of time before atomic power will be available at costs that make it commercially practicable. To go much beyond this statement on the basis of present information would be mere speculation. I view with some reserve the more enthusiastic predictions that the load of labor is to be lifted from the backs of men and that plenty will be enjoyed by all. The evidence for this optimism is not yet adequate.

Why speculate ahead of the evidence? Instead I call attention to the opportunity open for a collaborative study by physicists, engineers, industrial economists, and public administrators. The object of this study would not be to guess about the social adjustments which commercial atomic power might produce, but to follow closely scientific and engineering developments and describe, as the facts permit, the social issues and adjustments which are likely to ensue. Usually the interpretation of such change comes after the event. There is here an opportunity for natural scientists and social scientists, within certain limits, to predict social change and thereby to facilitate intelligent adjustment to it.

POLITICS AND GOVERNMENT

The current tendency in discussions of social effects of atomic energy is to emphasize the dangers of war and the possibilities of commercial power. The political and moral implications of atomic power are usually ignored. But in the future these implications may be found to be at least as important.

On the political side atomic energy looms as a giant new force propelling us towards the organization of society from the center. I give three examples: The May-Johnson Bill is one; and, although I share the concern which many scientists express over that bill, I suspect that any bill regulating atomic energy, which even the critics of the May-Johnson Bill might write, would contain many more restrictions on atomic physicists than have been traditional in scientific work. For the second example, consider the same centralizing tendency which would probably operate in the field of atomic power. When atomic power is developed for commercial purposes, it will not be developed by private utilities, or even by municipally owned or state owned systems. The power will be developed in Federal central stations, and the system will be a Federal system. Although existing channels of distributing the power may be used, the policies for production and distribution will be increasingly Federal policies. To the extent to which atomic power supplements other forms of power, the importance of central policies and central control will grow. The third example is in the field of political rights. If this country and Russia should go in for an armament race in atomic bombs, and the American people should come to realize that the Communist Party in this country is not only a political arm of the Soviet Union but could be a potentially important military arm, one wonders how long our traditional guarantees of political freedom would be maintained.

Such cases justify a query as to the impact, not only of atomic bombs and energy, but of all their born and unborn colleagues upon democratic political institutions. Moreover, what is the effect of such organizing of society from the center upon the status and independence of individuals; and, further upon the chances for peaceful international relations? How may the advantages of modern science and technology be reconciled with the conditions of effective political democracy and peaceful international relations? . . .

MORALS

Finally, I can only mention the moral implications of atomic energy. What is the effect on man of this possession of the ultimate power of the atom? It has been interesting to me to notice that, at the various conferences on atomic energy which I have attended, controls at the international level were discussed, and controls at the national level also; but the controls of a moral and educational nature within the individual were not mentioned even when representatives of theological seminaries were present. The individual, seemingly, had become too small change to matter. . . . The tragedy of the atomic bomb lies basically in the fact that, to survive in a world of such powers, we are compelled towards even greater emphasis and preoccupation with the means and the instrumentalities and the psychology of power. (It is significant to me that thoughtful people in this country were more possessed by fear at the time of Hiroshima, when we were at the zenith of our power, than they were at Pearl Harbor when our defenses were down.)

There will be much declamation and exhortation about these social-moral implications of atomic energy and other scientific inventions. But, ahead of declamation and exhortation, we need honest clarification. The minds which can really help society to understand the moral implications of modern scientific developments may be those of natural scientists (who are so deeply concerned with the product of their labors), or humanists, or men of affairs, or social scientists. To contribute helpfully these minds need (1) to be really capable of "the general view," (2) to understand the realities of the social world and the history of social ideas.

In conclusion, I return to my figure of the iceberg. It may be that further knowledge about atomic energy and further development of science and technology will point to problems of social adjustment quite different from those which I have stressed. But my guess is that, whatever the character of future scientific development, the main questions will continue to be questions of impact on war and peace, on power, on politics, and on morals. Whatever the course of developments, atomic bombs have made clear that the present is a time for greatness—a time for the abatement of pettiness of spirit—and a time for the pooling of disciplines.

I conclude with a quip which some wag used: "Atomic energy is here to stay but are we?"

HEADLINE COMICS

ATOMIC MAN, 1946

Comic books provided early examples of American culture wrestling with the impact of nuclear energy. *Atomic Man* appeared in March 1946—the first in a long tradition of comic book characters inspired by nuclear energy that includes Spider-Man, the X-Men, and many others. In the six-issue run of this atomic superhero, military veteran Adam Mann struggles with the consequences of an accident that concentrated the power of an atomic bomb in his right hand, seeking to control his newfound power and to use it for good. What cultural anxieties about nuclear energy does this comic book reveal?

Headline Comics, *Atomic Man* (March–April 1946), #18, available at Digital Comic Museum, http://digitalcomicmuseum.com.

ATOMIC MAN
To discharged veteran Adam Mann, research worker, comes the power of atomic energy— through an accident unique in the annals of chemistry— aware of the potency of this power- Adam strives to control it...
At the laboratory where Adam Mann works...
As long as I wear this lead glove over my right hand where the shrapnel in my wound is impregnated with radio activity, it is dormant—but why can't I control it? Perhaps, by will power, I can hold back some of its energy...

ARTHUR H. COMPTON

"THE ATOMIC CRUSADE AND ITS SOCIAL IMPLICATIONS," 1947

Many people believed that nuclear energy would radically transform American society—even if no one was quite sure exactly how. Speculation on the social and cultural impact of the new technology became a frequent subject of popular discussion. Some commentators—like Nobel Prize–winning physicist Arthur H. Compton—envisioned a world of leisure, peace, and freedom supported by the abundant energy captured from the atom. Others imagined a dystopic future of centralized state authority and the breakdown of the moral order. Compton spent World War II working with the Manhattan Project, supervising the nuclear reactors that converted uranium to plutonium for use in the first atomic bombs. In this essay he suggests some of the reasons that physicists worked on the project, and he envisions a future in which nuclear energy transforms human life for the better. What does Compton see as the promise of nuclear energy? Who does he think will benefit from it, and how?

ATOMIC POWER IN PEACE

It would be a mistake to suppose that either the scientists or the Government set out initially to build an atomic bomb. This was indeed the

---

Excerpted from Arthur H. Compton, "Atomic Crusade and Its Social Implications," *Annals of the American Academy of Political and Social Sciences* 249 (January 1947): 9–19. Reprinted by permission of SAGE Publications, Inc.

central military objective of the great atomic war effort. The bomb was, however, only the wartime aspect of a much greater vision. This vision began to take shape with the discovery fifty years ago that within the atom lies a storehouse filled with energy vaster by far than that which shows itself in such chemical processes as the burning of coal. Many a physicist in his heart of hearts hoped that he might have a share in presenting this wealth of energy in useful form as a Promethean gift to mankind. Perhaps nothing that physics could ever do would be of so great practical importance.

Dreams were dreamed of a more abundant life of greater knowledge to control disease, of greater freedom to build a better world. When uranium fission was discovered, it seemed that these dreams might be made real. Atomic power to drive the wheels of industry? Yes, and to propel ships over the seas and supply heat in the arctic wilds, making more of the planet available to man.

But uranium fission came at a time when war, the defense of all that was dear, compelled everyone's attention. The possibility of atomic explosions had been thought of only as terrors to be avoided, disasters that might overcome the bold experimenters who first would start the atomic chain reaction. Could atomic engines win the war? Hardly. By the time they were developed in usable form the war should be over. Nor would the use of such engines be of decisive importance. But the sudden release of atomic energy might make a bomb that would give to its user an enormous advantage. When this advantage was clearly seen, fear lest the enemy might first build such weapons called for a great effort. The atomic war program quickly took shape and into it was thrown all the strength that could be spared from other vital tasks.

To those who had been working with atoms for years, however, even the winning of the war was only one step in the use of the new-found strength. Victory was necessary so that people should be free to work for a better world. Among the essential features of that better world stands prominently the freedom from fear of war. The atomists knew that from here on, war would be so destructive that its waging would be madness. The world must see that this is true and be compelled to find a way whereby war can be prevented. With this as a greater objective, the years they spent at making atomic bombs prepared those who were making them to burst into a vast missionary call for peace as soon as the war was won. The little group of atomic physicists had now grown to a crusading army, with the strength of the many thousands of humanity-minded men and women who had shared their war effort.

Nor is peace itself the final goal. Many have been the frustrations of science. Improved methods of supplying food and shelter and other essentials to needy humanity have failed to achieve their promise because of the failure of society to use them for the common welfare. Here in atomic energy is a new, great opportunity to enrich life. Those who have brought this new child of science into being are determined that they shall not be frustrated again. It is not the rich, not the clever or the powerful, not the United States, Canada, or Britain alone that shall prosper from this new gift. The whole world shall have peace and, as far as the new advances of science and technology can bring it, prosperity and a more complete life. It is this great goal that the atomists hold before them. Atomic energy gives perhaps the greatest opportunity they will ever have to work effectively toward that goal. This opportunity must be used to the utmost. Such is the spirit of the atomic crusade. . . .

One of the unprecedented features of the atomic program was the large-scale co-operation of academic scientists, industrialists, and military men. The original ideas and their reduction to laboratory practice were primarily the contribution of men from the universities. The huge production program called for the strength of our best-organized industries. The undertaking of transferring the academic knowledge to industrial production and military use could in wartime be co-ordinated only through the military. . . .

The enforced wartime association of scientists and engineers has had at least some effect in giving each of these professions a greater appreciation of the task the other is performing. What is more, hundreds of young men have learned by practical experience the art of engineering research. It remains to be seen whether our universities and technical schools will be able to develop a postwar education that will supply the new men so urgently needed for solving the new technical problems of industry.

LASTING EFFECTS OF THE ATOMIC CRUSADE

The great human significance of the atomic crusade is its dramatic emphasis on the vital necessity of co-operation versus antagonism, of intelligent versus emotional or unconsidered living, and of great, commonly accepted objectives. These needs have become increasingly evident with every advance of science and technology. They are, in fact, essential to civilization. But the atomic bomb has written them before us in blazing letters.

The atomic bomb has made any future war between nations armed with such weapons so disastrous to both parties as to be irrational. The only way to avoid such disaster is that of international co-operation, intelligently planned, to prevent any nation from initiating a war. Only by progressive elimination of antagonisms and inculcation of a desire for co-operation can we hope to attain the great objective of a long-enduring peace with freedom.

Science and technology have brought with them increasing specialization. Management and labor, the various trades and professions, government with its many branches, business and agriculture, school and church, each is developing toward doing a better job in a narrower field. This system greatly increases our strength and richness of life as long as we work effectively together. When antagonisms develop, however, as in a nation-wide strike or an unwillingness to give justice to a minority group, the complex modern society becomes weakened so that all of us suffer. . . .

Above all, the atomic project is an example of the supreme value of a purpose. The war goal of the atomists was to build bombs that would bring victory and lasting peace. This was a part of the greater goal of defense and victory. The lesson one learns is that when people are working with a will to attain an objective, they will strive to learn how to do their part, and will willingly work with others as may be necessary for the desired result. With a goal established, training has a meaning and the will to co-operate is taken for granted. Here is the secret of co-ordinating the effort of free people.

In the success of the atomic bomb project the United States has perhaps caught a new view of its titanic strength. It is a strength that comes when a compelling objective draws the co-ordinated effort of trained and educated citizens. Those who have shared in the atomic crusade are now insisting that this titanic strength be turned toward ensuring the world's peace and toward giving the world's people a chance to share more fully in making their lives of value. Toward this great task the mobilization of atomic strength of the Nation was but a step.

H. M. PARKER

"SPECULATIONS ON LONG-RANGE WASTE DISPOSAL HAZARDS," 1948

In 1943 the U.S. Army established the Hanford Works, a top-secret facility for the production of plutonium, in southern Washington along the banks of the Columbia River. From the start, engineers and administrators at the Hanford Works struggled with questions of how to manage liquid and solid radioactive wastes generated by plutonium production. In 1944 physicist Herbert Parker began working at Hanford, where he became the facility's expert on radiation safety and waste management. Parker and his colleagues needed to devise both short- and long-term solutions to waste disposal that would not slow down the plant's ability to produce plutonium. What factors does Parker consider in assessing the disposal of radioactive wastes? How does he assess safety, environmental, and financial concerns? What assumptions does he make about technological improvement, and how do these assumptions condition his conclusions?

....

The operating policies of the Hanford Works call for the retention in buried tanks of all strongly radioactive wastes from the Separations Plants. The provisional studies during the war indicated that it was feasible to discharge less active wastes to ground to avoid absurd costs on

---

Excerpted from H. M. Parker, "Speculations on Long-Range Waste Disposal Hazards," January 26, 1948, U.S. Department of Energy's Public Reading Room Catalog, Document Number HW-8674, http://reading-room.labworks.org/Catalog/Default.aspx.

tank storage, evaporation equipment, or equivalent. For approximately one year the H. I. [Health Instrument] Section has studied the disposition of underground wastes for the following purposes:

(1) To determine whether there is significant risk of pollution of potable water sources, from the past practices,
(2) To devise corrective steps if such risk exists,
(3) To ascertain whether additional radioactive liquors can be similarly released, and thus to effect considerable economies.

ANTICIPATED DISPOSAL SITUATION

The salient facts from well drilling to date are:

(1) Plutonium contamination is very readily held by Hanford subsoils. The contamination was nowhere penetrated to a depth of more than 25 feet from the point of entry, although it has traveled laterally to about 200 feet.
(2) Fission product contamination is moderately well held by subsoil. It has penetrated downward as much as 100 feet in two years in one case. Lateral penetration up to nearly 300 feet has occurred.
(3) When the two species of contaminant are introduced together, they are separated out. Presence of fission products does not measurably alter the soil retention of plutonium.
(4) In the one special case of uranium waste disposal, rapid percolation to ground waters was noted. . . .

MAJOR DISASTERS

Suppose that *all* the buried tanks are disrupted, and the material carried immediately to the Columbia River. . . .

Common sense indicates that the postulated catastrophe is out of the question. Already enough is known about adsorption of the radioactivity in soil to ensure that almost all the activity would be held up locally. It would take years for the pollution to reach the river by underground routes. Ten years has been quoted as the travel time of underground water, and the contamination front should move more slowly than this. . . .

With 5 years or more for method development, it may be feasible to recover or divert a fraction of the escaped wastes, and thus alleviate the river pollution at the critical time. . . .

NECESSITY TO CONTINUE OR IMPROVE PRESENT PRACTICE

The above reasoning purports to show that there would be no critical hazard if, for example, all second cycle wastes were returned to the river through a filter bed. If so, one must defend the intention to build additional second cycle storage tanks, and consider also why even hotter wastes cannot be economically dumped into the ground. The answer is three-fold:

(1) A great time-wise extrapolation from existing data has been made.
(2) Conservative practice would never condone the *planned* pollution of a national asset such as the Columbia River with a long-lived contaminant such as plutonium. The contractor and the Atomic Energy Commission share a major responsibility in the proper protection of the public interest in this field. Once the river becomes contaminated with alpha-emitters, the present careful watch for beginning pollution from unsuspected sources will be impossible.
(3) If the rivers or water table sources were contaminated at any level approaching the presumed "tolerable" concentrations, an extensive survey and water sampling program would be needed over a very wide area. This would be expensive enough to represent a significant fraction of the cost saved on tanks. Coupled with the greatly lowered public morale in the Columbia Basin, it would be a poor business risk.

SUMMARY

Present disposal procedures may be continued, in the light of present knowledge, with the assurance of safety for a period of perhaps 50 years. Projection of the problem to future geological ages, as proposed by some authors, appears to be irrelevant in view of the technological progress in corrective measures that can be anticipated. Major *foreseeable* disasters would not seriously jeopardize the health of communities dependent on the river. In the worst case, radical curtailment of the use of river water may be required.

CONCLUSION

Currently planned geological studies are developing good data on the disposal system. While the continuation of these studies is deemed essential, there is no cause for hysteria, or for the radical expansion of the proposed program.

GENERAL ADVISORY COMMITTEE REPORTS ON BUILDING THE H-BOMB, 1949

One of the first major questions that American scientific, military, and political leaders faced in the earliest years of the Atomic Age was whether or not to develop the hydrogen, or thermonuclear, bomb. The "super bomb" contained truly devastating potential. Whereas the bomb dropped on Hiroshima delivered a 15 kiloton (thousands of tons of TNT) explosion, a thermonuclear blast might involve more than 15,000 kilotons. Some of the Manhattan Project scientists pursued initial research on the super bomb during the war. In 1949, after learning that the Soviets had detonated an atomic bomb, President Truman asked the Atomic Energy Commission's General Advisory Committee (GAC) for a recommendation on whether or not to pursue the super bomb. The GAC consisted of scientific and technical experts, most of whom had also participated in the Manhattan Project. In this document the GAC outlines several reasons *not* to pursue the new weapon. How did they reach this conclusion? To what extent did these scientists characterize the decision as a matter of science, a matter of politics, or a matter of ethics? How do the two opinions presented here differ from each other?

....

The General Advisory Committee has considered at great length the question of whether to pursue with high priority the development of the super bomb. No member of the Committee was willing to endorse

---

Excerpted from General Advisory Committee to the U.S. Atomic Energy Commission, October 30, 1949, www.atomicarchive.com/Docs/Hydrogen/GAC Report.shtml.

this proposal. The reasons for our views leading to this conclusion stem in large part from the technical nature of the super and of the work necessary to establish it as a weapon. . . .

It is notable that there appears to be no experimental approach short of actual test which will substantially add to our conviction that a given model will or will not work. . . . This does not mean that further theoretical studies would be without avail. It does mean that they could not be decisive. A final point that needs to be stressed is that many tests may be required before a workable model has been evolved or before it has been established beyond reasonable doubt that no such model can be evolved. Although we are not able to give a specific probability rating for any given model, we believe that an imaginative and concerted attack on the problem has a better than even chance of producing the weapon within five years.

A second characteristic of the super bomb is that once the problem of initiation has been solved, there is no limit to the explosive power of the bomb itself except that imposed by requirements of delivery. . . . Taking into account the probable limitations of carriers likely to be available for the delivery of such a weapon, it has generally been estimated that the weapon would have an explosive effect some hundreds of times that of present fission bombs. This would correspond to a damage area of the order of hundreds of square miles, to thermal radiation effects extending over a comparable area. . . . It needs to be borne in mind that for delivery by ship, submarine or other such carrier, the limitations here outlined no longer apply and that the weapon is from a technical point of view without limitations with regard to the damage that it can inflict.

It is clear that the use of this weapon would bring about the destruction of innumerable human lives; it is not a weapon which can be used exclusively for the destruction of material installations of military or semi-military purposes. Its use therefore carries much further than the atomic bomb itself the policy of exterminating civilian populations. It is of course true that super bombs which are not as big as those here contemplated could be made, provided the initiating mechanism works. In this case, however, there appears to be no chance of their being an economical alternative to the fission weapons themselves. It is clearly impossible with the vagueness of design and the uncertainty as to performance as we have them at present to give anything like a cost estimate of the super. If one uses the strict criteria of damage area per dollar and if one accepts the limitations on air carrier capacity likely to obtain in the years immediately ahead, it appears uncertain to us whether the super will be cheaper or more expensive that [*sic*] the fission bomb.

Although the members of the Advisory Committee are not unanimous in their proposals as to what should be done with regard to the super bomb, there are certain elements of unanimity among us. We all hope that by one means or another, the development of these weapons can be avoided. We are all reluctant to see the United States take the initiative in precipitating this development. We are all agreed that it would be wrong at the present moment to commit ourselves to an all-out effort toward its development.

We are somewhat divided as to the nature of the commitment not to develop the weapon. The majority feel that this should be an unqualified commitment. Others feel that it should be made conditional on the response of the Soviet government to a proposal to renounce such development. The Committee recommends that enough be declassified about the super bomb so that a public statement of policy can be made at this time. . . . It should explain that the weapon cannot be explored without developing it and proof-firing it. In one form or another, the statement should express our desire not to make this development. It should explain the scale and general nature of the destruction which its use would entail. It should make clear that there are no known or foreseen nonmilitary applications of this development. The separate views of the members of the Committee are attached to this report for your use.

J. R. Oppenheimer

MAJORITY ANNEX

We have been asked by the Commission whether or not they should immediately initiate an "all-out" effort to develop a weapon whose energy release is 100 to 1000 times greater and whose destructive power in terms of area of damage is 20 to 100 times greater than those of the present atomic bomb. We recommend strongly against such action.

We base our recommendation on our belief that the extreme dangers to mankind inherent in the proposal wholly outweigh any military advantage that could come from this development. Let it be clearly realized that this is a super weapon; it is in a totally different category from an atomic bomb. The reason for developing such super bombs would be to have the capacity to devastate a vast area with a single bomb. Its use would involve a decision to slaughter a vast number of civilians. We are alarmed as to the possible global effects of the radioactivity generated by the explosion of a few super bombs of conceivable magnitude. If super bombs will work at all, there is no inherent limit in the destructive power that may be attained with them. Therefore, a super bomb might become a weapon of genocide.

The existence of such a weapon in our armory would have far-reaching effects on world opinion; reasonable people the world over would realize that the existence of a weapon of this type whose power of destruction is essentially unlimited represents a threat to the future of the human race which is intolerable. Thus we believe that the psychological effect of the weapon in our hands would be adverse to our interest.

We believe a super bomb should never be produced. Mankind would be far better off not to have a demonstration of the feasibility of such a weapon, until the present climate of world opinion changes.

It is by no means certain that the weapon can be developed at all and by no means certain that the Russians will produce one within a decade. To the argument that the Russians may succeed in developing this weapon, we would reply that our undertaking it will not prove a deterrent to them. Should they use the weapon against us, reprisals by our large stock of atomic bombs would be comparably effective to the use of a super.

In determining not to proceed to develop the super bomb, we see a unique opportunity of providing by example some limitations on the totality of war and thus of limiting the fear and arousing the hopes of mankind.

James B. Conant; Hartley Rowe; Cyril Stanley Smith;
L. A. DuBridge; Oliver E. Buckley; J. R. Oppenheimer

MINORITY ANNEX: AN OPINION ON THE DEVELOPMENT OF THE "SUPER"

A decision on the proposal that an all-out effort be undertaken for the development of the "Super" cannot in our opinion be separated from consideration of broad national policy. A weapon like the "Super" is only an advantage when its energy release is from 100–1000 times greater than that of ordinary atomic bombs. The area of destruction therefore would run from 150 to approximately 1000 square miles or more.

Necessarily such a weapon goes far beyond any military objective and enters the range of very great natural catastrophes. By its very nature it cannot be confined to a military objective but becomes a weapon which in practical effect is almost one of genocide.

It is clear that the use of such a weapon cannot be justified on any ethical ground which gives a human being a certain individuality and dignity even if he happens to be a resident of an enemy country. It is evident to us that this would be the view of peoples in other countries. Its use would put the United States in a bad moral position relative to the peoples of the world.

Any postwar situation resulting from such a weapon would leave unresolvable enmities for generations. A desirable peace cannot come from such an inhuman application of force. The postwar problems would dwarf the problems which confront us at present.

The application of this weapon with the consequent great release of radioactivity would have results unforeseeable at present, but would certainly render large areas unfit for habitation for long periods of time.

The fact that no limits exist to the destructiveness of this weapon makes its very existence and the knowledge of its construction a danger to humanity as a whole. It is necessarily an evil thing considered in any light.

For these reasons we believe it important for the President of the United States to tell the American public, and the world, that we think it wrong on fundamental ethical principles to initiate a program of development of such a weapon. At the same time it would be appropriate to invite the nations of the world to join us in a solemn pledge not to proceed in the development or construction of weapons of this category. If such a pledge were accepted even without control machinery, it appears highly probable that an advanced stage of development leading to a test by another power could be detected by available physical means. Furthermore, we have our possession, in our stockpile of atomic bombs, the means for adequate "military" retaliation for the production or use of a "super."

E. Fermi; I. I. Rabi

LEWIS L. STRAUSS TO HARRY S. TRUMAN, 1949

Although the General Advisory Committee (GAC) recommended against the development of the super bomb, many members of the nuclear establishment disagreed with this recommendation. Government administrator, businessman, and naval officer Lewis L. Strauss served as one of the initial five commissioners of the AEC and eventually became chair of the commission. The secretaries of War and Defense also supported the development of the super bomb, and President Truman authorized the initiative in 1950. The AEC detonated the first H-bomb in 1954. How does Strauss address the ethical concerns raised by the scientists of the GAC in his letter to Truman? What does this document reveal about early Cold War rhetoric and attitudes about the Soviet Union? What were the implications of the logic laid out in the GAC report and in Strauss's letter for the arms race and for American encounters with nuclear energy?

I believe that the United States must be as completely armed as any possible enemy. From this, it follows that I believe it unwise to renounce, unilaterally, any weapon which an enemy can reasonably be expected to possess. I recommend that the President direct the Atomic Energy Commission to proceed with the development of the thermonuclear bomb, at highest priority subject only to the judgment of the Department of

Excerpted from Lewis L. Strauss to Harry S. Truman, November 25, 1949, National Security Archive, George Washington University, http://nsarchive.gwu.edu.

Defense as to its value as a weapon, and of the advice of the Department of State as to the diplomatic consequences of its unilateral renunciation or its possession. In the event that you may be interested, my reasoning is appended in a memorandum.

Lewis L. Strauss

PREMISES

(1) The production of such a weapon appears to be feasible (i.e., better than a 50–50 chance).
(2) Recent accomplishments by the Russians indicate that the production of a thermonuclear weapon is within their technical competence.
(3) A government of atheists is not likely to be dissuaded from producing the weapon on "moral" grounds. . . .
(5) The time in which the development of this weapon can be perfected is perhaps of the order of two years, so that a Russian enterprise started some years ago may be well along to completion.
(6) It is the historic policy of the United States not to have its forces less well armed than those of any other country (viz., the 5:3:3 naval ratio, etc., etc.)
(7) Unlike the atomic bomb which has certain limitations, the proposed weapon may be tactically employed against a mobilized army over an area of the size ordinarily occupied by such a force.
(8) The Commission's letter of November 9th to the President mentioned the "possibility that the radioactivity released by a small number (perhaps ten) of these bombs would pollute the earth's atmosphere to a dangerous extent." Studies requested by the Commission have since indicated that the number of such weapons necessary to pollute the earth's atmosphere would run into many hundreds. Atmospheric pollution is a consequence of present atomic bombs if used in quantity.

CONCLUSIONS

(1) The danger in the weapon does not reside in its physical nature but in human behavior. Its unilateral renunciation by the United States

could very easily result in its unilateral possession by the Soviet Government. I am unable to see any satisfaction in that prospect.

(2) The Atomic Energy Commission is competent to advise the President with respect to the feasibility of making the weapon; its economy in fissionable material as compared with atomic bombs; the possible time factor involved; and a description of its characteristics as compared with atomic bombs. Judgment, however, as to its strategic or tactical importance for the armed forces should be furnished by the Department of Defense, and views as to the effects on friendly nations or unilateral renunciation of the weapon is a subject for the Department of State. My opinion as an individual, however, based upon discussion with military experts is to the effect that the weapon may be critically useful against a large enemy force both as a weapon of offense and as a defensive measure to prevent landings on our own shores.

(3) I am impressed with the arguments which have been made to the effect that this is a weapon of mass destruction on an unprecedented scale. So, however, was the atomic bomb when it was first envisaged and when the National Academy of Sciences in its report of November 6, 1941, referred to it as "of superlatively destructive power." . . .

(4) Obviously the current atomic bomb as well as the proposed thermonuclear weapon are horrible to contemplate. All war is horrible. Until, however, some means is found of eliminating war, I cannot agree with those of my colleagues who feel that an announcement should be made by the President to the effect that the development of the thermonuclear weapon will not be undertaken by the United States at this time. This is because: (a) I do not think the statement will be credited in the Kremlin; (b) that when and if it should be decided subsequent to such a statement to proceed with the production of the thermonuclear bomb, it might in a delicate situation, be regarded as an affirmative statement of hostile intent; and (c) because primarily until disarmament is universal, our arsenal must not be less well equipped than with the most potent weapons that our technology can devise.

PART 2

BUILDING CONSENSUS

The patterns of thought that defined American responses to nuclear energy hardened in the 1950s. As the Cold War intensified, U.S. diplomatic leaders committed to the policy of containment—an attempt to stop the global spread of communism. From a military perspective this meant a commitment to nuclear weapons as a permanent and prominent part of the American arsenal. In the early 1950s the Eisenhower administration announced the "New Look" military strategy, which suggested that reliance on nuclear weapons and a willingness to use them provided the best and most efficient counter to the larger size of the Soviet Union's conventional forces. The ideas of massive retaliation and deterrence bolstered this new perspective—the very potential of a nuclear response might deter aggressive action by the Soviet Union. This logic propelled the arms race for the next four decades.

During this period, Americans also envisioned the ways that nuclear energy could revolutionize the economy. Nuclear technologies had the potential to make energy cheaper, food safer, and consumer goods more durable. Nuclear energy thus advanced Cold War aims—that is, a strong economy would check Soviet expansion and reaffirm the democratic and economic principles of American society. These elements combined to form the postwar nuclear consensus. How do the documents in part 2 express this consensus? What kinds of connections do these documents reveal between ideas about affluence, citizenship, and progress? What assumptions do they make about the economic potential and environmental impacts of nuclear energy?

"NATIONAL SECURITY COUNCIL RESOLUTION 68," 1950

In 1950, President Harry Truman asked his National Security Council (NSC) to reevaluate American geopolitical strategy in light of the reports that the Soviet Union had detonated an atomic bomb. The resulting top-secret document, known as NSC-68, outlined the policy of containment and defined American strategy during the first two decades of the Cold War. The NSC suggested a rationale for this policy that tied together the need for a strong standing military with nuclear capabilities, American ideals of freedom and justice, and an expanding capitalistic economic system. The NSC-68 thus provided the basic building blocks of the nuclear consensus. Why did the national security advisers believe that this was necessary? What kinds of military, economic, and social decisions could be justified with the logic laid out in this document? What alternatives might the NSC have considered?

. . . .

Two complex sets of factors have now basically altered this historic distribution of power. First, the defeat of Germany and Japan and the decline of the British and French Empires have interacted with the development of the United States and the Soviet Union in such a way that power has increasingly gravitated to these two centers. Second, the Soviet Union, unlike previous aspirants to hegemony, is animated by a new fanatic faith, antithetical to our own, and seeks to impose its absolute authority over the rest of the world. Conflict has, therefore, become endemic and is waged, on the part of the Soviet Union,

---

Excerpted from "National Security Council Resolution 68, April 14, 1950," *Naval War College Review* 27 (May–June 1975): 51–108.

by violent or non-violent methods in accordance with the dictates of expediency. With the development of increasingly terrifying weapons of mass destruction, every individual faces the ever-present possibility of annihilation should the conflict enter the phase of total war.

On the one hand, the people of the world yearn for relief from the anxiety arising from the risk of atomic war. On the other hand, any substantial further extension of the area under the domination of the Kremlin would raise the possibility that no coalition adequate to confront the Kremlin with greater strength could be assembled. It is in this context that this Republic and its citizens in the ascendancy of their strength stand in their deepest peril.

The issues that face us are momentous, involving the fulfillment or destruction not only of this Republic but of civilization itself. They are issues which will not await our deliberations. With conscience and resolution this Government and the people it represents must now take new and fateful decisions.

FUNDAMENTAL PURPOSE OF THE UNITED STATES

The fundamental purpose of the United States is laid down in the Preamble to the Constitution: "... to form a more perfect Union, establish Justice, insure domestic Tranquility, provide for the common defence, promote the general Welfare, and secure the Blessings of Liberty to ourselves and our Posterity." In essence, the fundamental purpose is to assure the integrity and vitality of our free society, which is founded upon the dignity and worth of the individual.

Three realities emerge as a consequence of this purpose: Our determination to maintain the essential elements of individual freedom, as set forth in the Constitution and Bill of Rights; our determination to create conditions under which our free and democratic system can live and prosper; and our determination to fight if necessary to defend our way of life, for which as in the Declaration of Independence, "with a firm reliance on the protection of Divine Providence, we mutually pledge to each other our lives, our Fortunes and our sacred Honor."

FUNDAMENTAL DESIGN OF THE KREMLIN

The fundamental design of those who control the Soviet Union and the international communist movement is to retain and solidify their absolute power, first in the Soviet Union and second in the areas now under their control. In the minds of the Soviet leaders, however, achievement of this design requires the dynamic extension of their authority and the ultimate elimination of any effective opposition to their authority.

The design, therefore, calls for the complete subversion or forcible destruction of the machinery of government and structure of society in the countries of the non-Soviet world and their replacement by an apparatus and structure subservient to and controlled from the Kremlin. To that end Soviet efforts are now directed toward the domination of the Eurasian land mass. The United States, as the principal center of power in the non-Soviet world and the bulwark of opposition to Soviet expansion, is the principal enemy whose integrity and vitality must be subverted or destroyed by one means or another if the Kremlin is to achieve its fundamental design.

THE UNDERLYING CONFLICT IN THE REALM OF IDEAS AND VALUES BETWEEN THE U.S. PURPOSE AND THE KREMLIN DESIGN

Nature of Conflict

The Kremlin regards the United States as the only major threat to the achievement of its fundamental design. There is a basic conflict between the idea of freedom under a government of laws, and the idea of slavery under the grim oligarchy of the Kremlin . . . and the exclusive possession of atomic weapons by the two protagonists. The idea of freedom, moreover, is peculiarly and intolerably subversive of the idea of slavery. But the converse is not true. The implacable purpose of the slave state to eliminate the challenge of freedom has placed the two great powers at opposite poles. It is this fact which gives the present polarization of power the quality of crisis.

The free society values the individual as an end in himself, requiring of him only that measure of self discipline and self restraint which make the rights of each individual compatible with the rights of every other individual. The freedom of the individual has as its counterpart, therefore, the negative responsibility of the individual not to exercise his freedom in ways inconsistent with the freedom of other individuals and the positive responsibility to make constructive use of his freedom in the building of a just society. . . .

The Soviet Union is developing the military capacity to support its design for world domination. The Soviet Union actually possesses armed forces far in excess of those necessary to defend its national territory. These armed forces are probably not yet considered by the Soviet Union to be sufficient to initiate a war which would involve the United States. This excessive strength, coupled now with an atomic capability, provides the Soviet Union with great coercive power for use in time of peace in furtherance of its objectives and serves as a deterrent to the victims of its aggression from taking any action in opposition to its tactics which would risk war. . . .

Our overall policy at the present time may be described as one designed to foster a world environment in which the American system can survive and flourish. It therefore rejects the concept of isolation and affirms the necessity of our positive participation in the world community.

This broad intention embraces two subsidiary policies. One is a policy which we would probably pursue even if there were no Soviet threat. It is a policy of attempting to develop a healthy international community. The other is the policy of "containing" the Soviet system. . . .

The United States now possesses the greatest military potential of any single nation in the world. The military weaknesses of the United States vis-a-vis the Soviet Union, however, include its numerical inferiority in forces in being and in total manpower. Coupled with the inferiority of forces in being, the United States also lacks tenable positions from which to employ its forces in event of war and munitions power in being and readily available. . . .

In the event of a general war with the U.S.S.R., it must be anticipated that atomic weapons will be used by each side in the manner it deems best suited to accomplish its objectives. In view of our vulnerability to Soviet atomic attack, it has been argued that we might wish to hold our atomic weapons only for retaliation against prior use by the U.S.S.R. To be able to do so and still have hope of achieving our objectives, the non-atomic military capabilities of ourselves and our allies would have to be fully developed and the political weaknesses of the Soviet Union fully exploited. In the event of war, however, we could not be sure that we could move toward the attainment of these objectives without the U.S.S.R.'s resorting sooner or later to the use of its atomic weapons. Only if we had overwhelming atomic superiority and obtained command of the air might the U.S.S.R. be deterred from employing its atomic weapons as we progressed toward the attainment of our objectives. . . .

It appears to follow from the above that we should produce and stockpile thermonuclear weapons in the event they prove feasible and would add significantly to our net capability. Not enough is yet known of their potentialities to warrant a judgment at this time regarding their use in war to attain our objectives. . . .

A more rapid build-up of political, economic, and military strength and thereby of confidence in the free world than is now contemplated is the only course which is consistent with progress toward achieving our fundamental purpose. The frustration of the Kremlin design requires the free world to develop a successfully functioning political and economic system and a vigorous political offensive against the Soviet Union. These, in turn, require an adequate military shield under which they can develop. It is necessary to have the military power to deter, if possible, Soviet expansion, and to defeat, if necessary, aggressive Soviet or Soviet-directed actions of a limited or total character. The potential strength of the free world is great; its ability to develop these military capabilities and its will to resist Soviet expansion will be determined by the wisdom and will with which it undertakes to meet its political and economic problems. . . .

The execution of such a build-up, however, requires that the United States have an affirmative program beyond the solely defensive one of countering the threat posed by the Soviet Union. This program must light the path to peace and order among nations in a system based on freedom and justice, as contemplated in the Charter of the United Nations. Further, it must envisage the political and economic measures with which and the military shield behind which the free world can work to frustrate the Kremlin design by the strategy of the cold war; for every consideration of devotion to our fundamental values and to our national security demands that we achieve our objectives by the strategy of the cold war, building up our military strength in order that it may not have to be used. The only sure victory lies in the frustration of the Kremlin design by the steady development of the moral and material strength of the free world and its projection into the Soviet world in such a way as to bring about an internal change in the Soviet system. Such a positive program—harmonious with our fundamental national purpose and our objectives—is necessary if we are to regain and retain the initiative and to win and hold the necessary popular support and cooperation in the United States and the rest of the free world.

FEDERAL CIVIL DEFENSE ADMINISTRATION

THIS IS CIVIL DEFENSE, 1951

The intensifying Cold War prompted a wide range of cultural and social responses. As the arms race accelerated, the prospect of a nuclear confrontation between the United States and the Soviet Union became easier to envision. The Federal Civil Defense Administration (FCDA), created in 1950, oversaw the development of a response program designed to instruct citizens on what to do in the case of a nuclear attack. The civil defense program largely consisted of organizing neighborhoods into teams with discrete responsibilities to respond to a nuclear attack or other disaster. Scholars have found in the materials produced by the FCDA a treasure trove for examining American ideas about citizenship, conformity, gender roles, and economic behavior in the early 1950s, even while they have questioned how seriously citizens engaged with the civil defense program.

This pamphlet introduced citizens to the idea of civil defense. What vision for the responsibilities of citizenship is laid out in this document? How did these expectations for citizenship intersect with the social context of the early 1950s, a time known for its virulent anticommunism, social homogeneity, and economic prosperity? How did the civil defense program relate to the developing nuclear consensus?

• •

Modern civil defense is nothing like civil defense in previous wars. Where once our danger was from fire bombs and high explosives, now it

---

Excerpted from Federal Civil Defense Administration, *This Is Civil Defense* (Washington, D.C.: U.S. Government Printing Office, 1951), 3–8, 10.

is from the atomic bomb. The wide oceans that used to protect us have given way to the global bomber. Today we face more kinds of attack than ever before, and our danger is much greater.

This booklet was written to give you, as a responsible American citizen, the straight facts on why civil defense is needed, how it works, and what part you must play to make it a success.

WHAT IS CIVIL DEFENSE?

Civil defense is a way of saving lives and property. It is a way of protecting you and your family in case of war on the United States. It is a way of helping to keep you going, and to keep production going, in spite of atomic, biological or chemical attacks.

One of the chief aims of civil defense is to help you to stay at work no matter what may come. Unless all of us kept at our jobs in the face of attack, the enemy would win the war. His aim would be to make you and others quit and desert your cities so that our defense plants would shut down. *Your* aim would be to keep working and to give our armed forces the things they need to beat the enemy.

CAN AMERICA BE ATTACKED?

Yes. At any time.

Right now enemy planes can reach every major city in the United States. We know that Russia has heavy, long-range bombers patterned after our own B-29. Most of these bombers could get through our defenses if an attack came.

We know Russia has atomic bombs, and is making more all the time. . . .

Atomic bombs could be delivered by enemy aircraft. So could disease or poison gas. They could be delivered at any moment.

What is more, fifth columnists within our own country could strike at us with all three types of weapons. Saboteurs and enemy aircraft could attack at the same time. Or, saboteurs could start their work much earlier. With some types of biological warfare, they could begin work weeks or even months ahead of time, without waiting for a war to start.

CAN WE DEFEND OURSELVES?

There is no known way of preventing most enemy bombers from reaching their targets in the United States.

Gen. Hoyt Vandenberg, Chief of Staff of the United States Air Force, has said that *at most* we could knock down only 30 out of each 100 enemy planes attacking the United States. At least 7 out of 10 would get through. . . .

You can be sure that everything possible will be done to stop the enemy at our borders, and to stop sabotage before it starts within our borders. But you also can be sure that, in case of war, a good percentage of enemy attacks would be successful in spite of all we could do.

That is why we must have civil defense—and have it now!

HOW WOULD CIVIL DEFENSE HELP?

There is a good defense against modern weapons and civil defense is it.

We cannot prevent enemy attacks from happening—but we can keep them from knocking us out of the war. If we know what to do we can save lives and property, restore our cities and get back into the fight no matter what form those attacks might take.

There is no sure way of keeping enemy planes from getting through—but there is a sure way to save many thousands of lives if we are attacked. Civil defense is the way, and the *only* way.

There are good defenses against the atomic bomb. There are ways to save thousands of people from the worst effects of blast, heat, and radioactivity. There are ways to take shelter, and to rescue the trapped and injured, and to cut fire losses to a minimum. Civil defense can show you what they are. . . .

Civil defense services can be organized to bring in help from outside, and to get a stricken city back into working order in the shortest possible time. Civil defense can provide food, shelter and medical care for the victims of enemy attack.

Civil defense can make it possible for you to stick with your job, your home, and your family no matter what might come. It can make staying in your city safer than trying to get out of it.

But we must face facts. Civil defense takes planning, organization and a lot of hard work. It would not protect every life and every home. However, with civil defense, *most* people and *most* homes could be saved. . . .

WHO IS RESPONSIBLE FOR CIVIL DEFENSE?

You are.

Civil defense is set up by Federal and State law. But no law in the world will work unless you back it up by your own actions. That's why, in the end, the responsibility for civil defense is yours.

The thing to remember is this: If the bombs from enemy planes ever fell on your city, they would not fall on a plan, or an organization, or a system of government. They would fall on you and your family and friends.

If you were a soldier, you would be trained to take care of yourself and keep on fighting. As a defender of the home front, you must learn to protect yourself and keep on working. Despite every precaution, a soldier might be killed. So might you. But the more you know, and the better trained you are, the better your chances for survival.

The whole idea of civil defense is to help you protect yourself, and to make the best use of your own special ability and skill in an emergency. Then you will be able to save yourself and others if trouble comes.

FEDERAL CIVIL DEFENSE ADMINISTRATION

WOMEN IN CIVIL DEFENSE, 1952

The federal civil defense effort strongly reinforced mainstream ideas about gender roles in the 1950s. The Federal Civil Defense Administration (FCDA) dispersed its message in publications like this one, in which the agency tried to speak directly to American women. Schools provided another important medium for the civil defense initiative, and the FCDA produced booklets, pamphlets, movies, and classroom activity sheets that reinforced its messages of preparedness, patriotism, and individual responsibility. All of these documents helped define normative behaviors for American families in the 1950s.

What role for women does this document envision—in civil defense but also in broader society? In what ways did the conservative gender roles of the 1950s relate to the other defining elements of the nuclear consensus?

• •

The home is the basic unit of the community—and the basic unit on which defense of the home front must be built.

Whether you are a housewife, secretary, business executive, or nurse, civil defense looks to you, as a woman, to take an active role in protecting your home. No one else can do that job for you.

Your first duty in civil defense is to act at once to educate your family

---

Excerpted from Federal Civil Defense Administration, *Women in Civil Defense* (Washington, D.C.: Federal Civil Defense Administration, 1952), 1–6.

in self-protection against modern weapons, and to make your home as safe as possible against the dangers of enemy attack.

Your second duty is to participate in your community civil defense organization. There must be a basic civil defense organization in each community in the United States, regardless of size or location. Without fully organized communities, there can be no adequate national civil defense program.

If your community does not have an active civil defense organization, much of the blame must fall on you and your neighbors. Unless you, as a responsible American woman, take action you are gambling with the safety of your family, your friends, your community, and your country.

You would hardly blame others for failing to provide food, clothing, and shelter for your family. That is your family responsibility. And so is family civil defense. Community civil defense can be effective only if the families of the community are solidly behind it, willing to give time and effort to make it work. National civil defense can be only as effective as the people of the Nation make it. . . .

Furthermore, an enemy would strike at our cities and our people first. This is true because our two greatest strengths are the civilian will to fight and to produce the sinews of war. To win a war, military forces must have a constant pipeline of supplies flowing to the fighting fronts. Civilians produce the things the military forces need. If our people, farms, and factories are destroyed, the military forces will soon have no supplies with which to fight.

And remember that American soldiers, sailors, airmen, and marines are fighting for the people at home. If the home front crumples behind them, they not only have nothing to fight *with*, they have nothing to fight *for*.

That's why civil defense is just as important as a strong military force and why civil defense is important to every community and person in America. Not every community will be attacked. But those that are attacked cannot hope to take care of themselves without help. The help must come from the organized civil defense forces of communities and States which are not attacked. That's why *all* communities must be organized.

To do the job, over 17 million hard-working, well-trained volunteers are needed. Your community needs volunteers now and for years to come.

The greater percentage of these volunteers will be women like you.

At least 60 percent of civil defense volunteers must be women serving in hundreds of specialized civil defense jobs. Many volunteer jobs can be filled by you and your friends right now, with only a little training.

When you have trained your family and prepared your home, you have more than doubled your chances for survival in an atomic attack.

When you have joined in organizing your community, you have given the community and the Nation a far better chance to survive an enemy attack.

But you will have done more than just prepare in case of war—you will have made a positive contribution to keeping the peace.

An unprepared nation invites attack. A nation without civil defense is unprepared.

A strong civil defense preparedness program, like a strong military preparedness program, is not just a shield but a sword. Adequate civil defense preparedness can actually help hold the enemy at bay. If the enemy knows that he can demoralize us by an all-out attack on the homefront; if he knows that we are not prepared for it; if he knows that our civil defense system is ill-manned, ill-trained, and ill-equipped—this is a direct invitation to launch such an attack on our people and on our cities.

But, if Russia knows that millions of American men and women are well trained and organized and ready to move into action when the attack comes; if Russia knows that we have thousands of trained rescue squads and tens of thousands of wardens and millions of American families trained in first aid and self-protection; if Russia knows that we have this kind of adequate civil defense preparedness which would save at least half the American lives that might otherwise be lost—then Russia, or any other enemy, will think long and hard before launching an attack on this country. . . .

Said quite simply, a strong American civil defense program forces Russia to use two atomic bombs instead of one, thus reducing the size and effectiveness of the Russian stockpile.

A strong civil defense stands side by side with our armed forces as a major deterrent to enemy attack on our own country. This makes civil defense a major force in helping keep the peace and in preventing World War III. . . .

Getting America prepared on the home front is a responsibility that falls in large part on the shoulders of all American women. It's your job—and you have no time to waste. . . .

SETTING UP A HOME SHELTER

You should provide a shelter area for your family immediately. In most houses, a portion of the basement is the best available area.

A suitable shelter space can be set up with little or no construction or expense. Before selecting your shelter area, ask the advice of your local civil defense warden. He can pass on to you technical advice from the local civil defense engineering services.

In selecting and preparing your shelter, follow these general principles:

1. There should be minimum danger from flying glass, falling beams, and debris.
2. There should be two outside exits on different sides of your shelter area. If there are not, take every precaution against the possibility of debris blocking the single exit. The shelter should be well ventilated.
3. The shelter should be stocked with a three-day supply of water and canned food in sealed containers, first-aid supplies . . . a flashlight with extra batteries, and other emergency necessities.

If you live in a place other than your own home, find out the location of the nearest shelter area.

FIRST AID AND HOME NURSING

Everyone should take the Red Cross standard first-aid course. In addition to its value as a civil defense self protection measure, a knowledge of first aid is especially important for housewives, because of the many accidents that occur at home. Training you can receive in a Red Cross home nursing course is also very valuable, as it teaches you to recognize symptoms of illness and to carry out the physician's orders. . . .

FIRE PREVENTION AND FIRE FIGHTING

Fire is one of the greatest dangers from enemy air attack. Start today to prepare your home against this danger. Firemen say "a clean building seldom burns." Trash piles, rubbish, or stored odds and ends that accumulate around the home increase the danger of fire. Good housekeeping is the first line of defense against fire. . . .

If an enemy bombing attack is made on your city, it will start more fires than the fire department can handle. Fighting fires in the home or neighborhood will be up to each family; if such fires are not put out, they can join to burn up whole areas. Thus a knowledge of fire-fighting techniques in the home is a necessity. You should know what tools to have and how to use them.

DWIGHT D. EISENHOWER

"ADDRESS BEFORE THE GENERAL ASSEMBLY OF THE UNITED NATIONS ON PEACEFUL USES OF ATOMIC ENERGY," 1953

President Dwight D. Eisenhower delivered the "Atoms for Peace" speech before the United Nations in 1953. He sought to change the rhetoric of nuclear energy from one of fear to one of hope by focusing on the economic potential of nuclear energy both at home and abroad. The president proposed a system of international control of nuclear energy that would allow countries to develop peaceful nuclear technologies as long as they renounced the use of the atom for war. How does Eisenhower portray the impact of atomic energy on everyday life? In what ways does he stake out a position relative to the Soviet Union, and how might the Cold War context have shaped his motives and vision? How does this public speech compare to a classified document like National Security Council Report 68 (NSC-68)?

I feel impelled to speak today in a language that in a sense is new—one which I, who have spent so much of my life in the military profession, would have preferred never to use.

That new language is the language of atomic warfare.

Excerpted from Dwight D. Eisenhower, "Address before the General Assembly of the United Nations on Peaceful Uses of Atomic Energy, New York City," in *Public Papers of the Presidents of the United States: Dwight D. Eisenhower. Containing the Public Messages, Speeches, and Statements of the President: January 20 to December 31, 1953* (Washington, D.C.: Government Printing Office, 1960), 813–22.

The atomic age has moved forward at such a pace that every citizen of the world should have some comprehension, at least in comparative terms, of the extent of this development of the utmost significance to every one of us. Clearly, if the peoples of the world are to conduct an intelligent search for peace, they must be armed with the significant facts of today's existence.

My recital of atomic danger and power is necessarily stated in United States terms, for these are the only incontrovertible facts that I know. I need hardly point out to this Assembly, however, that this subject is global, not merely national in character.

On July 16, 1945, the United States set off the world's first atomic explosion. Since that date in 1945, the United States of America has conducted 42 test explosions.

Atomic bombs today are more than 25 times as powerful as the weapons with which the atomic age dawned, while hydrogen weapons are in the ranges of millions of tons of TNT equivalent. . . .

If at one time the United States possessed what might have been called a monopoly of atomic power, that monopoly ceased to exist several years ago. . . .

The free world, at least dimly aware of these facts, has naturally embarked on a large program of warning and defense systems. That program will be accelerated and expanded.

But let no one think that the expenditure of vast sums for weapons and systems of defense can guarantee absolute safety for the cities and citizens of any nation. The awful arithmetic of the atomic bomb does not permit of any such easy solution. Even against the most powerful defense, an aggressor in possession of the effective minimum number of atomic bombs for a surprise attack could probably place a sufficient number of his bombs on the chosen targets to cause hideous damage.

Should such an atomic attack be launched against the United States, our reactions would be swift and resolute. But for me to say that the defense capabilities of the United States are such that they could inflict terrible losses upon an aggressor—for me to say that the retaliation capabilities of the United States are so great that such an aggressor's land would be laid waste—all this, while fact, is not the true expression of the purpose and the hope of the United States.

To pause there would be to confirm the hopeless finality of a belief that two atomic colossi are doomed malevolently to eye each other indefinitely across a trembling world. To stop there would be to accept helplessly the probability of civilization destroyed—the annihilation of the irreplaceable heritage of mankind handed down to us generation from generation—and the condemnation of mankind to begin all over again the age-old struggle upward from savagery toward decency, and right, and justice.

Surely no sane member of the human race could discover victory in such desolation. Could anyone wish his name to be coupled by history with such human degradation and destruction.

Occasional pages of history do record the faces of the "Great Destroyers" but the whole book of history reveals mankind's never-ending quest for peace, and mankind's God-given capacity to build.

It is with the book of history, and not with isolated pages, that the United States will ever wish to be identified. My country wants to be constructive, not destructive. It wants agreements, not wars, among nations. It wants itself to live in freedom, and in the confidence that the people of every other nation enjoy equally the right of choosing their own way of life.

So my country's purpose is to help us move out of the dark chamber of horrors into the light, to find a way by which the minds of men, the hopes of men, the souls of men everywhere, can move forward toward peace and happiness and well being. . . .

The United States, heeding the suggestion of the General Assembly of the United Nations, is instantly prepared to meet privately with such other countries as may be "principally involved," to seek "an acceptable solution" to the atomic armaments race which overshadows not only the peace, but the very life, of the world.

We shall carry into these private or diplomatic talks a new conception.

The United States would seek more than the mere reduction or elimination of atomic materials for military purposes.

It is not enough to take this weapon out of the hands of the soldiers. It must be put into the hands of those who will know how to strip its military casing and adapt it to the arts of peace.

The United States knows that if the fearful trend of atomic military buildup can be reversed, this greatest of destructive forces can be developed into a great boon, for the benefit of all mankind.

The United States knows that peaceful power from atomic energy is no dream of the future. That capability, already proved, is here—now—today. Who can doubt, if the entire body of the world's scientists and engineers had adequate amounts of fissionable material with which to test and develop their ideas, that this capability would rapidly be transformed into universal, efficient, and economic usage.

To hasten the day when fear of the atom will begin to disappear from the minds of people, and the governments of the East and West, there are certain steps that can be taken now.

I therefore make the following proposals:

The Governments principally involved, to the extent permitted by elementary prudence, to begin now and continue to make joint contributions from their stockpiles of normal uranium and fissionable materials to an International Atomic Energy Agency. We would expect that such an agency would be set up under the aegis of the United Nations. . . .

The Atomic Energy Agency could be made responsible for the impounding, storage, and protection of the contributed fissionable and other materials. The ingenuity of our scientists will provide special safe conditions under which such a bank of fissionable material can be made essentially immune to surprise seizure.

The more important responsibility of this Atomic Energy Agency would be to devise methods whereby this fissionable material would be allocated to serve the peaceful pursuits of mankind. Experts would be mobilized to apply atomic energy to the needs of agriculture, medicine, and other peaceful activities. A special purpose would be to provide abundant electrical energy in the power-starved areas of the world. Thus the contributing powers would be dedicating some of their strength to serve the needs rather than the fears of mankind.

The United States would be more than willing—it would be proud to take up with others "principally involved" the development of plans whereby such peaceful use of atomic energy would be expedited.

Of those "principally involved" the Soviet Union must, of course, be one.

I would be prepared to submit to the Congress of the United States, and with every expectation of approval, any such plan that would:

First—encourage world-wide investigation into the most effective peacetime uses of fissionable material, and with the certainty that they had all the material needed for the conduct of all experiments that were appropriate;

Second—begin to diminish the potential destructive power of the world's atomic stockpiles;

Third—allow all peoples of all nations to see that, in this enlightened age, the great powers of the earth, both of the East and of the West, are interested in human aspirations first, rather than in building up the armaments of war;

Fourth—open up a new channel for peaceful discussion, and initiate at least a new approach to the many difficult problems that must be solved in both private and public conversations, if the world is to shake off the inertia imposed by fear, and is to make positive progress toward peace.

Against the dark background of the atomic bomb, the United States does not wish merely to present strength, but also the desire and the hope for peace.

The coming months will be fraught with fateful decisions. In this Assembly; in the capitals and military headquarters of the world; in the hearts of men everywhere, be they governors or governed, may they be the decisions which will lead this world out of fear and into peace.

To the making of these fateful decisions, the United States pledges before you—and therefore before the world—its determination to help solve the fearful atomic dilemma—to devote its entire heart and mind to find the way by which the miraculous inventiveness of man shall not be dedicated to his death, but consecrated to his life.

UNION CARBIDE AND CARBON CORPORATION

"WHAT DOES ATOMIC ENERGY REALLY MEAN TO YOU?" 1953

This 1953 advertisement in *Fortune Magazine* for the Union Carbide and Carbon Corporation promised a "future unlimited" and advancements in "medicine, agriculture, and industry." Union Carbide ran the Oak Ridge National Laboratory, one of the Atomic Energy Commission's primary research facilities, and is considered one of the founding companies of the modern petrochemical industry. What messages about nuclear energy, progress, and nature does the image convey?

Union Carbide and Carbon Corporation advertisement, "What does Atomic Energy really mean to you?" *Fortune Magazine*, May 1953. Courtesy of the Union Carbide Corporation.

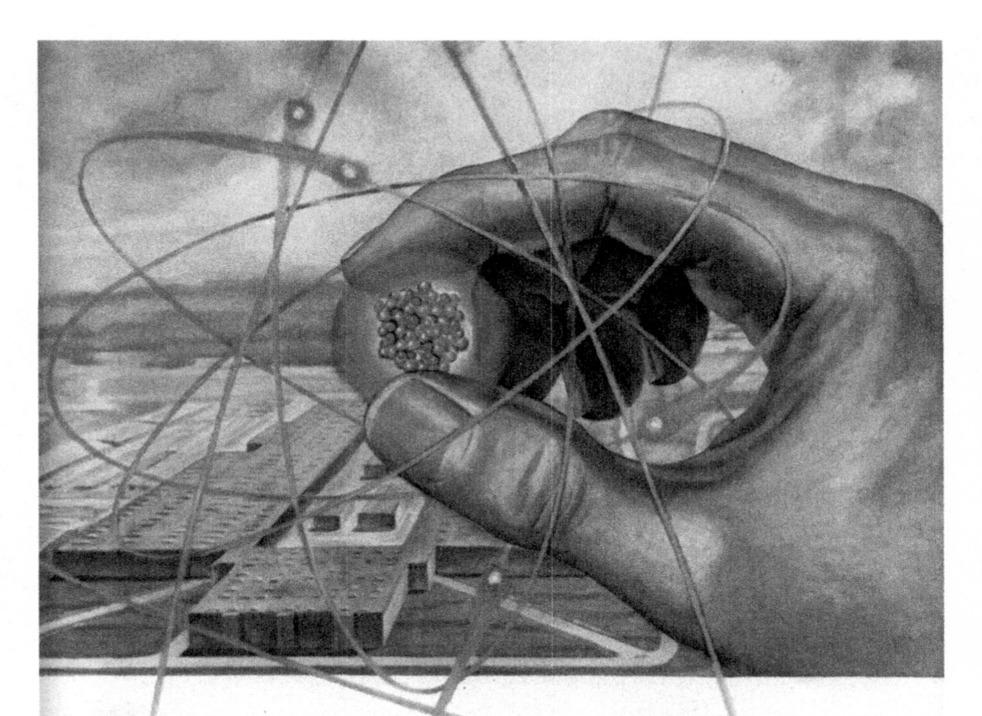

What does Atomic Energy really mean to you?

Dramatic new developments in medicine, agriculture, and industry promise long-time benefits for us all

Scientists have long known that the secret core of the atom concealed vast stores of concentrated energy. Evidence that man had unlocked the secret came with the atomic bomb.

Then came the task of developing methods to release this unbounded energy slowly, gradually, in ways of lasting benefit to all of us.

ISOTOPES AN EXAMPLE—When uranium atoms are split they emit a barrage of highly active particles. Certain chemicals placed in this barrage become radioactive and shoot off particles from themselves. Substances thus treated are called radioactive isotopes.

When these chemicals are made radioactive their paths can be traced through plants and animals, showing the organs they affect. This may increase our understanding of the processes of life itself.

FUTURE UNLIMITED—Atomic energy is also proving useful in industrial research and production, such as analyzing metals and other materials. It promises to be even more valuable, however, in providing concentrated power for transportation, home, and industry.

UNION CARBIDE'S PART—From the beginning UCC has had a hand in the mining and treatment of uranium ores, the development of engineering processes, and the production of special materials for the atomic energy program. Under Government contract Union Carbide manages and operates the huge research and production installations at Oak Ridge, Tenn. and Paducah, Ky.

All of this activity fits in with the continuing efforts of the people of Union Carbide to transform the elements of the earth into useful materials for science and industry.

FREE: *Learn more about the interesting things you use every day. Write for the illustrated booklet "Products and Processes" which tells how science and industry use the* ALLOYS, CARBONS, CHEMICALS, GASES, *and* PLASTICS *made by Union Carbide. Ask for booklet D.*

UNION CARBIDE
AND CARBON CORPORATION
30 EAST 42ND STREET UCC NEW YORK 17, N. Y.

UCC's Trade-marked Products of Alloys, Carbons, Chemicals, Gases, and Plastics include

SYNTHETIC ORGANIC CHEMICALS • EVEREADY Flashlights and Batteries • NATIONAL Carbons • ACHESON Electrodes • PYROFAX Gas
ELECTROMET Alloys and Metals • HAYNES STELLITE Alloys • PREST-O-LITE Acetylene
DYNEL TEXTILE FIBERS • BAKELITE, KRENE, and VINYLITE Plastics • LINDE Oxygen • PRESTONE and TREK Anti-Freezes

LEWIS L. STRAUSS

"MY FAITH IN THE ATOMIC FUTURE," 1955

In 1953, President Dwight Eisenhower appointed Lewis Strauss chair of the Atomic Energy Commission (AEC). Strauss is in some ways best remembered for a quotation that he did not actually issue: that nuclear fission would make commercial energy "too cheap to meter." Although he never said this, the misquotation aptly captures Strauss's belief in the technology's potential. As chair of the AEC, he helped develop the Atoms for Peace program introduced earlier in this section. Here, Strauss outlines the many uses of the peaceful atom. On what does he base his faith? What will it take, according to Strauss, for this type of atomic future to develop? What does this document reveal about Strauss's ideas about the role of government, business, and science?

. . . .

Many people regard the atomic discoveries of recent years as part of a nightmare that disrupts the peaceful dreams of civilized man. I do not believe history will see them in that light. We have gained control over natural forces that can advance civilization, even within a single generation, to a point which man has never attained before. I believe firmly that our knowledge of the atom is intended by the Creator for the service and not the destruction of mankind. The Atomic Energy Act of 1946 was a farsighted law. But I had certain specific reservations about it. Nuclear energy, which I believed could change the world, was

---

Excerpted from Lewis L. Strauss, "My Faith in the Atomic Future," *Reader's Digest* (August 1955): 17–21. Reprinted with permission from Reader's Digest.

straitjacketed in Government regulations. Research, development, patents, manufacturing and possession of fissionable materials were denied to private enterprise. Atomic energy was an absolute Government monopoly.

Atomic weapons development is necessarily a Government responsibility. But I was convinced that developments in agriculture, industry and power production would not be realized fully until the field was opened to the genius and enterprise of American industry.

Actually, the restrictions might have been relaxed sooner but for the attitude of Soviet Russia. Beginning in 1946, when the United States held a virtual world monopoly on nuclear weapons, we proposed international control, subject to rigid inspections and enforcement, which would have limited the use of atomic energy to peaceful purposes. At that time we even offered to share our knowledge and resources with all nations.

The Soviets did everything possible to delay, confuse and destroy that plan. Actually, they were launching their own secret atomic program. We detected their clandestine weapon test in 1949, and were at once engaged in the costly and perilous contest for supremacy in nuclear weapons. Every thinking person knows now that our present great and versatile stockpile is the major safeguard of the free world.

Meanwhile, atomic energy became associated in popular thinking with death and destruction. Yet the custodians of atomic energy under President Truman and President Eisenhower never lost sight of its benign potentials. Progress was phenomenal in the production of radioisotopes. They were produced by AEC reactors as early as 1947. They were distributed freely to institutions here and abroad, and within a few years revolutionized some areas of medical research and the diagnosis and treatment of certain diseases. Scarce, high-priced radium for the treatment of cancer was rendered virtually obsolete by radioactive cobalt and other elements which are equally effective sources of gamma rays and yet are now available to institutions at a small fraction of radium's cost.

Several different types of nuclear reactors for the generation of electrical power were designed by the AEC. But most authorities put the date of their construction in the remote future.

When I returned to the AEC as chairman in 1953 I was deeply impressed by the growing conviction in the White House and the

Congress that the time had come for full-scale development of atomic energy outside the military area.

President Eisenhower, in his address to the General Assembly of the United Nations on December 8, 1953, stated: "The United States pledges before the world its determination to help solve the fearful atomic dilemma, to devote its entire heart and mind to find the way by which the miraculous inventiveness of man shall not be dedicated to his death, but consecrated to his life."

Two months later the President sent to Congress the message which resulted in the Atomic Energy Act of 1954. The new law had two great aims—to make international cooperation possible, and to enable private enterprise to develop the atom for peaceful purposes.

The progress of the past 18 months—only a moment in history—has been extraordinary. For example, the AEC announced its program to develop power-producing reactors, and invited private companies to participate. The quick response was totally unexpected. The Duquesne Light Co. is building our first full-scale nuclear-power plant at Shippingport, Pa. At least four or five others will be constructed in the near future in Massachusetts, Michigan, Nebraska, Illinois and New York.

These pioneer nuclear-power plants cannot be economically competitive with conventional plants at present. Yet the participating companies are paying about 90 *percent* of the total costs! This, I maintain, could only happen under free enterprise in an expanding economy.

Indeed, two or three proposed plants will be constructed *entirely* without financial help from the Government. Mr. Hudson R. Searing, president of one of these companies, Consolidated Edison of New York, recently told stockholders that nuclear power "is the only way we can see of bringing about lower electricity costs over the long pull."

It is pointless to speculate on how soon nuclear power will be cheaper than power produced by falling water or the burning of coal or oil. We do know that our resources of fossil fuels are limited, and that coal and oil will be needed for functions which atomic energy cannot perform. We know that there is a great disparity in electricity costs between those areas where water power, coal and oil are plentiful and regions like New England where such resources are scarce or nonexistent. We also must remember that there are many countries which are not blessed with such abundant resources as our own. So to me the present question of "economic" nuclear power is academic. I believe that it will be available before long, and that it logically will be used first where it is needed most.

Nuclear power for the propulsion of ships and aircraft will also come sooner than is generally realized. Few people have grasped the significance of the *Nautilus* and her sister submarine, the *Seawolf.* With the feasibility and safety of the marine propulsion reactor established beyond doubt, the job now is up to the designers and builders of surface vessels. The time to begin is *now.* That was the thought behind the President's recent recommendation of a nuclear-powered merchant ship.

I am convinced that the radioactive isotopes will continue to be the wonders of the atomic age. Today, they are being used by many industries to control processes, detect flaws and test the durability or wearing quality of all sorts of materials. New uses for them are found every day.

Used as "tracers" or as radiation sources, these atomic particles can search out the innermost secrets of nature and give man greater control over his environment. For example, plant geneticists have already used radiation from isotopes to produce a new strain of rust-resistant oats, wilt-resistant tomato seedlings, and a peanut plant with 30 percent greater yield. These and similar developments will mean millions—perhaps billions—of dollars to farmers.

By incorporating small amounts of radiophosphorus in fertilizers, and then using instruments to trace the uptake from the soil through roots, stem, leaves and blossom, agricultural experts can now determine the right amount of fertilizer to use in the most economical manner, and at the proper time in the growing cycle.

For nearly a century science has tried in vain to solve the fundamental secret of photosynthesis, the process whereby nature traps solar energy in the green leaf and converts water and carbon dioxide into the sugars and starches on which all higher life subsists. Using radioactive carbon as the tracer, researchers today seem to be on the point of solving (and perhaps duplicating) this mysterious process. If successful, that achievement might lead to the synthetic production of basic foodstuffs from simple and abundant chemicals—the solution to the world's pressing food problems.

Since 1946 the American people have spent more than 12 billion dollars on atomic energy. We will probably continue spending about two billion a year. Most of this money is invested in our stockpile of nuclear weapons, which represents the security of the free world. We have no choice but to maintain that security—until the whole world joins us in arriving at a safe solution to the "atomic dilemma." I firmly believe that can be accomplished.

But our nuclear stockpile also represents a national resource of incalculable value. With nuclear weapons you can "beat swords into plowshares and spears into pruning hooks" even more realistically than the Scriptures envisioned. The material is immediately convertible to peaceful uses. That is what President Eisenhower had in mind when he told the United Nations that the weapon "must be put in the hands of those who know how to strip its military casing and adapt it to the arts of peace."

Young people have asked me if I sincerely think that we shall enjoy the benefits of the atom before the world is overtaken by the destructive power that is within man's grasp. With all my heart I can answer: Yes!

We are living in an era that seems designed to test the courage and faith of free men. Yet I do not believe that any great discovery of the atom's magnitude came from man's intelligence alone. A Higher Intelligence decided that man was ready to receive it. My faith tells me that the Creator did not intend man to evolve through the ages to this stage of civilization only now to devise something that would destroy life on this earth.

My old chief, former President Herbert Hoover, to whose Quaker convictions all warfare is revolting, listened to President Eisenhower's U.N. speech and said: "I pray it may be accepted by all the world." We pray that Divine Providence will guide men of all nations to grasp this opportunity to "shake off the inertia imposed by fear and make positive progress toward peace."

HEINZ HABER

THE WALT DISNEY STORY OF OUR FRIEND THE ATOM, 1956

Walt Disney holds a fascinating place in U.S. cultural history. Most remember him as the genius of entertainment who introduced Americans to Mickey Mouse and the Magic Kingdom. Disney was also an aggressive Cold Warrior—taking a strong stance against Soviet expansion abroad and an active role in the domestic anticommunist movement. The document below includes the foreword (written by Disney himself) and the prologue to *Our Friend the Atom*, published in conjunction with an hour-long, animated television documentary with the same title. Both the book and the television show sought to explain the basic science behind nuclear energy. Both of them sought to convince the American public that nuclear energy represented the correct path for U.S. development. Why is Disney producing this book? What does he hope to accomplish? What is the moral of the fable of the nuclear genie? What might it tell us about Disney's—and his audience's—views on science, technology, and progress?

. . . .

FOREWORD

Fiction often has a strange way of becoming fact. Not long ago we produced a motion picture based on the immortal tale *20,000 Leagues under*

Textual excerpts from *The Walt Disney Story of Our Friend the Atom* by Heinz Haber published by Golden Press © 1956 Disney. See pp. 10–21, 137, 149, 159.

the Sea, featuring the famous submarine "Nautilus." According to that story the craft was powered by a magic force.

Today the tale has come true. A modern namesake of the old fairy ship—the submarine "Nautilus" of the United States Navy—has become the world's first atom-powered ship. It is proof of the useful power of the atom that will drive the machines of our atomic age.

The atom is our future. It is a subject everyone wants to understand, and so we long had plans to tell the story of the atom. In fact, we considered it so important that we embarked on several *atomic projects.*

For one, we are planning to build a Hall of Science in the TOMORROWLAND section of DISNEYLAND where we will—among other things—put up an exhibit of atomic energy. . . .

With our *atomic projects* we found ourselves deep in the field of nuclear physics. Of course, we don't pretend to be scientists—we are story tellers. But we combine the tools of our trade with the knowledge of experts. We even created a new Science Department at the Studio to handle projects of this kind. The story of the atom was assigned to Dr. Heinz Haber, Chief Science Consultant of our Studio. He is the author of this book and he helped us in developing our motion picture.

The story of the atom is a fascinating tale of human quest for knowledge, a story of scientific adventure and success. Atomic science has borne many fruits, and the harnessing of the atom's power is only the spectacular end result. It came about through the work of many inspired men whose ideas formed a kind of chain reaction of thoughts. These men came from all civilized nations, and from all centuries as far back as 400 B.C.

Atomic science began as a positive, creative thought. It has created modern science with its many benefits for mankind. In this sense our book tries to make it clear to you that we can indeed look upon the atom as our friend.

Walt Disney

PROLOGUE

Deep in the tiny atom lies hidden a tremendous force. The force has entered the scene of our modern world as a most frightening power of destruction, more fearful and devastating than man ever thought possible.

We all know of the story of the military atom, and we all wish that it weren't true. For many obvious reasons it would be better if it weren't real, but just a rousing tale. It does have all the earmarks of a drama: a frightful terror which everyone knows exists, a sinister threat, mystery and secrecy. It's a perfect tale of horror!

But, fortunately, the story is not yet finished. So far, the atom is a superb villain. Its power of destruction is foremost in our minds. But the same power can be put to use for creation, for the welfare of all mankind.

What will eventually be done with the atom? It is up to us to give the story a happy ending. If we use atomic energy wisely, we can make a hero out of a villain.

This, then, is the story of the atom. It is a story with a straightforward plot and a simple moral—almost like a fable. In many ways the story of the atom suggests the famous tale from the *Arabian Nights*: "The Fisherman and the Genie." Perhaps this tale even hints at what lies in our atomic future. . . .

THE FISHERMAN AND THE GENIE

There once lived an aged Fisherman, who dwelt in poverty with his wife and three children. Each day he cast his net into the sea four times, and rested content with what it brought forth.

One day, after three vain casts, the old Fisherman drew in his net for the fourth time. He found it heavier than usual. Examining his catch, he found among the shells and seaweeds a small brazen vessel. On its leaden stopper was the ancient seal of King Solomon.

"A better catch than fish!" he exclaimed. "This jar I can sell. And who knows what thing of value it might contain?"

With his knife he pried out the stopper. Then, as he peered into the jar, smoke began to pour from it. He fell back in astonishment as the smoke rose in a great dark column and spread like an enormous mushroom between earth and sky. And his astonishment turned into terror as the smoke formed into a mighty Genie, with eyes blazing like torches and fiery smoke whirling about him like the simoom of the desert.

"Alas!" cried the old Fisherman, falling to his knees. "Spare me, O Genie. I am but a poor man, who has not offended thee!"

The Genie glared down on the trembling old man.

"Know," he thundered, "that because thou hast freed me, thou must die. For I am one of those condemned spirits who long ago disobeyed the word of King Solomon. In this brazen vessel he sealed me, and he commanded that it be cast into the sea, there to lie forever—or until some mortal should, by unlikely chance, bring up the vessel from the depths and set me free."

The old Fisherman listened in silent fear as the Genie's eyes flamed.

"For centuries," the great voice of the Genie continued, "I lay imprisoned deep in the sea, vowing to grant to my liberator any wish—even to make him master of all the wealth in the world, should he desire it. But no liberator came. At last, in my bitterness, I vowed that my liberator, who had delayed so long, should have no wish granted him—except how he should die. Thou, old man, art my liberator, and according to my solemn vow thou must die!"

"O," wailed the Fisherman, "why was I born to set thee free? Why did I cast this net and bring forth from the deeps this accursed vessel? Why must thou reward me with death?"

The fiery smoke swirled more swiftly about the Genie, and he gestured with impatience.

"Fisherman," he roared, "delay not, but choose how thou wilt die!"

The old Fisherman was terrified indeed. Yet in this moment of danger he was able to bestir his wits.

"O Genie," he begged, "if I must die, so be it. But first grant me this one wish. Thy great form did seem to come forth out of this little vessel, and yet I cannot believe it. Prove to me that one who is so mighty can indeed fit into such a little vessel."

The Genie towered above the little fisherman. His eyes blazed brighter.

"Old man," he thundered, "thou shalt see, before thy death, that nothing lies beyond my powers."

Swiftly the Genie dissolved into smoke, and the smoke funneled back into the little vessel.

Instantly the Fisherman leaped forward and thrust the leaden stopper, bearing the seal of King Solomon, into the jar.

"Now," he shouted to the imprisoned Genie, "choose how thou, in thy turn, wilt die! A prisoner thou art again, and back into the depths will I fling thee. All fishermen, and their children, and their children's children, shall be warned of the wicked Genie and forbidden ever to cast their nets here. And at the bottom of the sea shalt thou lie forevermore!"

The Genie's agitated voice sounded faintly through the brazen vessel.

"Stop, stop! Only set me free once more, and thou shalt live!"

The Fisherman raised the vessel to cast it into the waves. "O Genie," he said, "only when I cast thee back into the sea shall I be safe."

The voice in the little vessel grew frantic. "Fisherman, hear me! Live thou shalt, and richly! Restore my freedom and I vow, by Allah, to grant thee three wishes, to make thee rich and happy all thy days. Good Fisherman, hear my solemn vow!"

The old man had little heart for revenge, and he bethought himself of what a friendly Genie might do for his ragged, hungry family. The Genie continued to entreat him for mercy. And at last the Fisherman pried out the stopper.

Once more the smoke poured forth, and again the giant form of the Genie loomed against the sky. With a great kick, the Genie sent the brazen vessel spinning far out over the waves.

The old Fisherman trembled, fearing the worst. But the Genie turned toward him, and bowed his towering form, and spoke gently.

"Fear not," he said. "You heard my vow. O Fisherman, my master, name thy three wishes. . . ."

This fable tells of the age-old wish of man to be the master of a mighty servant that does his bidding. But to us it has a still deeper meaning: the story of the atom is like that tale; we ourselves are like that fisherman. For centuries we have been casting our nets into the sea of the great unknown in search of knowledge. Finally a catch was made: man found a tiny vessel, the atom, in which lies imprisoned a mighty force—atomic energy.

Like the fisherman, man marveled at his strange find and examined it closely for its value. He pried it open—split it in two. And as he did so a terrible force was released that threatened to kill with the most cruel forms of death: death from searing heat, from the forces of a fearful blast, or from subtly dangerous radiations.

And as it was to the fisherman, it is to us a great, an almost unbelievable marvel that such a tremendous force could dwell in such a tiny vessel.

Here we are, we fishermen, marveling and afraid, staring at the terrifying results of our curiosity. The fable, though, has a happy ending; perhaps our story can, too. Like the Fisherman we must bestir our wits. We have the scientific knowhow to turn the Genie's might into peaceful and useful channels. He must at our beckoning grant three wishes for the good of man. The fulfillment of these wishes can and will reshape our future lives. . . .

Our First Wish: Power

The coal and oil resources of our planet are dwindling, yet we need more and more power. The atomic Genie offers us an almost endless source of energy. For the growth of our civilization, therefore, our first wish shall be for: POWER! . . .

Our Second Wish: Food and Health

Mankind has long suffered from hunger and disease. The atomic Genie offers us a source of beneficial rays. These are magic tools of research which can, above all, help us to produce more food for the world and to promote the health of mankind. Our second wish, therefore, shall be for: FOOD AND HEALTH! . . .

The Third Wish: Peace

There is left to us the third and last wish. It is an important one that demands wisdom. If the last wish is unwise, then—as some of the old legends tell—all the wishes granted before may be lost.

The atomic Genie holds in his hands the powers of both creation and destruction. The world has reason to fear those powers of destruction. They could yet destroy civilization and much of humankind.

So our last wish should simply be for the atomic Genie to remain forever our friend!

It lies in our own hands to make wise use of the atomic treasures given to us. The magic power of atomic energy will soon begin to work for mankind throughout the world. It will grant the gifts of modern technology to even the most remote areas. It will give more food, better health—the many benefits of science—to everyone.

We still have much to learn. But the key to a peaceful atomic future lies in the spirit of the great thinkers of the past. From them we have inherited a great wealth of knowledge. Whatever benefits the atom brings us will come from that heritage. . . .

When these scientists created their theories and made their discoveries, they perhaps hardly foresaw that there would ever be widespread application of their work. They simply marveled at the world around them and deeply desired to know about Nature and her ways. That the results of their noble efforts could or even would ever be applied for destruction—this was farthest from their minds and hearts.

They gave us knowledge of the atom, and our last and most important wish will come true if we use the power of this knowledge in their spirit.

Then the atom will become truly our friend.

BUREAU OF PUBLIC ROADS

A PRELIMINARY REPORT ON HIGHWAY NEEDS FOR CIVIL DEFENSE, 1956

The Federal-Aid Highway Act of 1956—also called the National Interstate and Defense Highways Act—prompted some of the most sweeping economic and environmental changes of any law in American history. By authorizing a nationwide system of highways that connected the country's major cities, the law transformed the economy of commercial transportation, patterns of residential housing, American car culture, and countless urban, suburban, rural, and wilderness landscapes. This report on highway needs identifies the links between civil defense preparedness, industrial capacity, and transportation infrastructure. Although national security concerns were not the only motivation for the development of highways, they played an important role in winning public support for the massive undertaking. How did concerns about civil defense shape plans for highway development? One of the more interesting conclusions of this report concerns the difficulty of evacuating major cities in response to a nuclear attack. How, then, do the reports authors justify the expansion of the highway system? What other conclusions do they draw?

THE PROBLEM

Briefly expressed, the problem under consideration is one of survival of populations in urban centers if subjected to enemy attack. Three

---

Excerpted from Bureau of Public Roads, *A Preliminary Report on Highway Needs for Civil Defense* (Washington, D.C.: U.S. Department of Commerce, 1956), 2–3, 41.

significant factors have been instrumental in establishing the dimensions of the problem; evaluation of these factors is required in any rational analysis of the data. These factors involve the present and future development of:

1. Nuclear weapons with their enlarged area of destruction and radioactive fallout potentialities.
2. Means of delivery of the bomb by air, submarine or clandestine methods.
3. Means of defense against delivery of the bomb, with the related factor of expected warning time to large population centers for possible evacuation programs.

Consideration of these three items in relation to the civil defense problem of the 185 target areas under consideration in this study indicates that:

1. The potential area of severe bomb blast is such as to make target areas almost wholly subject to first degree effects with severe problems of radioactive fallout in certain adjoining areas.
2. Urban centers would be vulnerable to attack on a total war basis, with greater vulnerability along the coastal areas.
3. Warning time would be short. Estimates released by the Federal Civil Defense Administration indicate from 1½ hours warning at border and coastal cities to 3 or 4 hours for cities in the interior of the country. The probability exists that developments in methods of delivery and detection have altered or may soon alter this factor of warning time, either increasing or decreasing it.

PURPOSE OF THE STUDY

In consideration of the conditions presently constituting the broad civil defense problem, it is the purpose of this study to determine on a nationwide basis:

1. The adequacy of the present highway and street systems in rural and suburban areas to permit evacuation of urban populations by motor vehicles.
2. The extent that roadway improvements in these areas would permit

faster or more complete evacuation of urban populations, and the estimated cost of such improvements.

3. The relationship between normal highway construction programs and increase in roadway capacity for evacuation purposes. . . .

In measuring the capabilities of the present road systems, the most significant finding of the study is the fact that approximately 32 million people of the 90.7 million in the 185 target areas could possibly be evacuated to a 15-mile distance from a selected ground zero in 1½ hours by motor vehicle under conditions that would place the roads under full capacity operations within a reasonably short time after warning. This, of course, would be dependent upon full preconditioning of the population for evacuation and efficient traffic operations. Equally important is the finding that total evacuation of most large urban areas over present street and highway systems within reasonable warning time is impossible. This does not imply that large numbers of people could not be removed from the target areas, but it does prove that when warning times are short additional means of survival must be considered for these places. In addition, evacuation to this distance would in no manner insure the civilian population against the possible effects of radio-active fallout. . . .

The civil defense problems of the nation are intimately related to the present road systems and to the expansion and development of these systems in the future. Looking at the longer range problems, the accelerated program of highway and street construction, if maintained, should provide a sound base for increasing the potentialities for possible survival in the atomic age in addition to the great benefits derived from normal economic growth and development of one of the most important national assets in deterring aggression—industrial capacity.

WALTER REUTHER

ATOMS FOR PEACE: A SEPARATE OPINION, 1956

By the mid-1950s a strong consensus had emerged about the importance of nuclear weapons as a linchpin of American geopolitical strategy and the potential of nuclear power to support the country's growing economy. Much less consensus existed, however, on how to develop nuclear power for commercial purposes. Should the development of nuclear power be a public or private endeavor? How might the development of a commercial nuclear power system be regulated? To help answer these questions, the Joint Committee on Atomic Energy (the congressional committee that oversaw the Atomic Energy Commission and other elements of the U.S. nuclear program) in 1956 empaneled a committee of industry, labor, and government leaders to report on the progress of the Atoms for Peace program. Walter Reuther, president of the United Auto Workers and one of the nation's most prominent labor leaders, sat on this panel.

In this document Reuther lays out his perspective on the role the federal government should play in promoting commercial nuclear power. Why does he think a strong role for the federal government is necessary? What does he see as the reasons for developing nuclear power? How should it be developed, and by whom? How does this document relate to the emerging nuclear consensus?

Excerpted from Walter Reuther, *Atoms for Peace: A Separate Opinion on Certain Aspects of the Report of the Panel on the Peaceful Uses of Atomic Energy to the Joint Congressional Committee on Atomic Energy* (Detroit: United Auto Workers, 1956), 1–5. Courtesy of the United Auto Workers.

In the cold war—in freedom's struggle against the forces of communist tyranny—in the struggle for the hearts and minds of men—speed, all possible speed, in harnessing the atom to man's peacetime needs can be decisive.

Access to low-cost nuclear power may prove the key to the economic development of backward areas and make possible the liberation of millions of people from poverty, hunger, ignorance and disease. America's leadership is essential if we are to block the communists in their efforts to forge poverty into power.

Our success in harnessing the atom to lift the burden of poverty and disease from hundreds of millions of the world's people living in hunger and ill-health would establish America in a position of moral leadership against which communist propaganda would be impotent.

Harnessing of the atom for peaceful purposes will give us the tools with which to wage freedom's most effective propaganda to these people—the propaganda of the democratic deed. Failure on the part of America to pursue the peaceful harnessing of the atom with maximum speed, determination and dedication may prove to be the Achilles' heel of the cold war.

PEACEFUL ATOM PROGRAM LAGS—I

We shall not give leadership to other people if we refuse to exercise it in our own behalf. The fact is that the United States is failing to demonstrate the outstanding leadership in releasing atomic energy for peaceful purposes which it demonstrated in putting the atom to work for war.

We are not moving with speed and determination to convert atomic energy into an instrument of peaceful progress. Our program for developing atomic energy as a source of electric power is moving much too slowly.

For many years after the war no really significant beginning was made to apply the atom to peaceful uses. Finally, one year ago, AEC invited private enterprise to submit proposals for participating in the development of atomic reactors for the generation of electric power.

But no private power reactors are now under construction and none has completed the initial stages of design.

The one large-scale reactor now building is the AEC demonstration reactor at Shippingport, Pa. Apart from this government project, the sobering fact is that, today, ten and a half years after the end of the war, America's peacetime atomic power program has not advanced beyond the drawing boards. The head of the AEC reactor division states that as of today there is no certainty when, if ever, private industry will build and operate a power reactor. . . .

NUCLEAR POWER ESSENTIAL TO ECONOMIC GROWTH—I

The need to develop atomic energy as a practical source of power for use in the United States is urgent. There are power-hungry areas in our country today. There are other areas where the high cost of power retards economic progress and is encouraging the flight of industry to other parts of the country.

Total power requirements in the United States will expand at a tremendous rate over the next 25 years. We shall need nuclear power to meet those requirements. I cannot accept the comfortable assurance that our conventional fuel resources will meet all our power needs for another 20 to 25 years. Nor will I rely upon the Federal Power Commission's consistently conservative forecasts of power requirements as reflecting the true growth potential of our economy or the increasing needs of the American people.

No power ceiling should be imposed upon the normal and necessary expansion of our economy. Of that we must make sure. We must develop every source of energy we have, including atomic energy.

HOW BEST TO GET THE JOB DONE—I

To meet the challenge and to realize the opportunity of the peaceful uses of atomic energy, we must mobilize all our forces and enlist the active participation of every segment of our economy. We must make full use of the capabilities both of government and of private enterprise. Only by drawing upon the special contributions of each can we make satisfactory progress toward our objectives—fortifying the strength of our nation, advancing the welfare of our people and discharging our world responsibilities.

The technological barriers ahead of us are formidable. Enormous investments are required. The financial risks are great. But all these difficulties can be overcome by a united, determined and co-ordinated effort.

It would be tragic to destroy this great opportunity for national achievement and world leadership by dissipating our strength in ideological warfare over the respective roles of government and private enterprise. That is a sure-fire formula for standing still here in the United States while the rest of the world moves forward in the practical application of atomic energy to human needs.

The opportunities for government and private enterprise to make their special contributions in this new field will arise out of the particular problems encountered at each stage of development. Both will have a vital part to play throughout, but the character and degree of their responsibilities will change as we make progress in mastering the new technology.

The early research and experimental phases of the program are primarily the government's responsibility, it is generally agreed. The development and construction of small scale power reactors is also primarily a government responsibility. . . . As to the next stage of the program, I question whether the present AEC policy of placing prime responsibility on private utility companies to provide risk capital for the construction and testing of full-scale demonstration reactors is sound. I share the point of view that the productive know-how and managerial skill of American private enterprise can make an important contribution to this phase of development, providing, however, the government takes the initiative and assumes the financial risks involved in the construction and testing of these full-scale reactors. Building these first full-scale power reactors is an extension of the research and experimental work which only the government is capable of performing under present circumstances. Accordingly, I suggest that the AEC policy of placing primary reliance on private enterprise at this stage of development be subjected to early review and reappraisal by the Joint Committee.

PROTECTING THE PUBLIC INTEREST FROM MONOPOLY—I

Once the practical possibilities of atomic energy have been demonstrated, a realistic and attractive opportunity will have been created for private enterprise to engage in the new atomic industries and develop their full potential. When this stage is reached, the government is obligated to guard against monopoly control in the new industries, and to make sure that consumers are protected by effective competition in the sale of nuclear power and other atomic services. Both publicly and privately financed electric utility systems should engage in supplying nuclear power to the public just as in the past they have supplied power generated from conventional energy sources.

PART 3

CHALLENGING CONSENSUS

Most Americans found the nuclear consensus difficult to challenge. Indeed, limited avenues for dissent helped define American society in the early 1950s, during the height of Senator Joseph McCarthy's search for Communist infiltration. But by the middle of the decade a growing awareness of the consequences of nuclear warfare and anxiety over its prospect fueled challenges to the nuclear consensus. In 1954 the U.S. conducted a thermonuclear test at Bikini Atoll in the Pacific Ocean—known as the Castle Bravo Test. At 15 kilotons the blast proved far more powerful than expected, and it exposed the populations of two small Pacific island communities to fallout. A Japanese fishing vessel, the *Lucky Dragon*, had also sailed into the blast area. Several days after the test, the boat returned to Japan, its twenty-three-person crew suffering from radiation exposure. Earlier nuclear tests had received little news coverage, but the Bravo Test captured media attention and sparked concern about the environmental and human health impacts of the arms race. The Cuban Missile Crisis in October 1962 brought the world to the brink of war and nuclear fear to an apex. This, too, prompted a rethinking of the nuclear consensus and some of the first steps away from the arms race.

The documents in part 3 explore the ways that some Americans adhered to the nuclear consensus even while the challenge to it gathered steam. An environmental claim lay at the heart of this challenge: nuclear technologies were causing unforeseen and unpredictable environmental consequences. How did the prospect of nuclear war change the way people thought about the environment? What kinds of human impacts on nature did nuclear energy allow? How did challenging the nuclear consensus cause Americans to rethink the meaning of citizenship (to the nation and to the species) and the shape of civil society?

BERTRAND RUSSELL AND ALBERT EINSTEIN

"THE RUSSELL-EINSTEIN MANIFESTO," 1955

For the first ten years of the Atomic Age, few Americans questioned the nuclear consensus—the benefits of the surging economy were spread widely, and dissent proved difficult in a climate of homogeneity. Growing concerns about the arms race and its social and environmental consequences, however, prompted the first significant challenges to the consensus. In 1955 eleven of the twentieth century's leading scientists and intellectuals issued what became known as "The Russell-Einstein Manifesto," named for philosopher Bertrand Russell and physicist Albert Einstein. In the manifesto, they responded, in particular, to the development and testing of the hydrogen bomb and the *Lucky Dragon* accident of 1954. The manifesto built on the alarm caused by the Bravo Test and called for a rethinking of the arms race. How did the signatories to the manifesto understand this incident? How do they appeal to Cold War rhetoric and try to transcend it? Who is the manifesto written to and for? What do the signers envision as the role of science and scientists in the arms race and opposition to it?

. . . .

In the tragic situation which confronts humanity, we feel that scientists should assemble in conference to appraise the perils that have arisen as a result of the development of weapons of mass destruction, and to discuss a resolution in the spirit of the appended draft.

Excerpted from "The Russell-Einstein Manifesto," July 9, 1955, Pugwash Conferences on Science and World Affairs, http://pugwash.org/1955/07/09/statement manifesto/. Courtesy of the Bertrand Russell Peace Foundation.

We are speaking on this occasion, not as members of this or that nation, continent, or creed, but as human beings, members of the species Man, whose continued existence is in doubt. The world is full of conflicts; and, overshadowing all minor conflicts, the titanic struggle between Communism and anti-Communism.

Almost everybody who is politically conscious has strong feelings about one or more of these issues; but we want you, if you can, to set aside such feelings and consider yourselves only as members of a biological species which has had a remarkable history, and whose disappearance none of us can desire.

We shall try to say no single word which should appeal to one group rather than to another. All, equally, are in peril, and, if the peril is understood, there is hope that they may collectively avert it.

We have to learn to think in a new way. We have to learn to ask ourselves, not what steps can be taken to give military victory to whatever group we prefer, for there no longer are such steps; the question we have to ask ourselves is: what steps can be taken to prevent a military contest of which the issue must be disastrous to all parties?

The general public, and even many men in positions of authority, have not realized what would be involved in a war with nuclear bombs. The general public still thinks in terms of the obliteration of cities. It is understood that the new bombs are more powerful than the old, and that, while one A-bomb could obliterate Hiroshima, one H-bomb could obliterate the largest cities, such as London, New York, and Moscow.

No doubt in an H-bomb war great cities would be obliterated. But this is one of the minor disasters that would have to be faced. If everybody in London, New York, and Moscow were exterminated, the world might, in the course of a few centuries, recover from the blow. But we now know, especially since the Bikini test, that nuclear bombs can gradually spread destruction over a very much wider area than had been supposed.

It is stated on very good authority that a bomb can now be manufactured which will be 2,500 times as powerful as that which destroyed Hiroshima. Such a bomb, if exploded near the ground or under water, sends radio-active particles into the upper air. They sink gradually and reach the surface of the earth in the form of a deadly dust or rain. It was this dust which infected the Japanese fishermen and their catch of fish.

No one knows how widely such lethal radio-active particles might be diffused, but the best authorities are unanimous in saying that a war with H-bombs might possibly put an end to the human race. It is feared that if many H-bombs are used there will be universal death—sudden only for a minority, but for the majority a slow torture of disease and disintegration.

Many warnings have been uttered by eminent men of science and by authorities in military strategy. None of them will say that the worst results are certain. What they do say is that these results are possible, and no one can be sure that they will not be realized. We have not yet found that the views of experts on this question depend in any degree upon their politics or prejudices. They depend only, so far as our researches have revealed, upon the extent of the particular expert's knowledge. We have found that the men who know most are the most gloomy.

Here, then, is the problem which we present to you, stark and dreadful and inescapable: Shall we put an end to the human race; or shall mankind renounce war? People will not face this alternative because it is so difficult to abolish war.

The abolition of war will demand distasteful limitations of national sovereignty. But what perhaps impedes understanding of the situation more than anything else is that the term "mankind" feels vague and abstract. People scarcely realize in imagination that the danger is to themselves and their children and their grandchildren, and not only to a dimly apprehended humanity. They can scarcely bring themselves to grasp that they, individually, and those whom they love are in imminent danger of perishing agonizingly. And so they hope that perhaps war may be allowed to continue provided modern weapons are prohibited.

This hope is illusory. Whatever agreements not to use H-bombs had been reached in time of peace, they would no longer be considered binding in time of war, and both sides would set to work to manufacture H-bombs as soon as war broke out, for, if one side manufactured the bombs and the other did not, the side that manufactured them would inevitably be victorious.

Although an agreement to renounce nuclear weapons as part of a general reduction of armaments would not afford an ultimate solution, it would serve certain important purposes. First, any agreement between East and West is to the good in so far as it tends to diminish tension. Second, the abolition of thermo-nuclear weapons, if each side believed that the other had carried it out sincerely, would lessen the fear of a sudden attack in the style of Pearl Harbour, which at present keeps both sides in a state of nervous apprehension. We should, therefore, welcome such an agreement, though only as a first step.

Most of us are not neutral in feeling, but, as human beings, we have to remember that, if the issues between East and West are to be decided in any manner that can give any possible satisfaction to anybody, whether Communist or anti-Communist, whether Asian or European or American, whether White or Black, then these issues must not be decided by war. We should wish this to be understood, both in the East and in the West.

There lies before us, if we choose, continual progress in happiness, knowledge, and wisdom. Shall we, instead, choose death, because we cannot forget our quarrels? We appeal as human beings to human beings: Remember your humanity, and forget the rest. If you can do so, the way lies open to a new Paradise; if you cannot, there lies before you the risk of universal death.

RESOLUTION

We invite this Congress, and through it the scientists of the world and the general public, to subscribe to the following resolution:

"In view of the fact that in any future world war nuclear weapons will certainly be employed, and that such weapons threaten the continued existence of mankind, we urge the Governments of the world to realize, and to acknowledge publicly, that their purpose cannot be furthered by a world war, and we urge them, consequently, to find peaceful means for the settlement of all matters of dispute between them."

Max Born, Percy W. Bridgman, Albert Einstein,
Leopold Infeld, Frederic Joliot-Curie, Herman J. Muller,
Linus Pauling, Cecil F. Powell, Joseph Rotblat,
Bertrand Russell, Hideki Yukawa

ROGER REVELLE AND MILNER B. SCHAEFER

"GENERAL CONSIDERATIONS CONCERNING THE OCEAN AS A RECEPTACLE FOR ARTIFICIALLY RADIOACTIVE MATERIALS," 1957

In the 1950s scientists struggled to figure out the environmental impacts of nuclear energy. The pursuit of information about such topics as fallout, radioactivity, and environmental and human health had dramatic consequences for the scientific understanding of natural processes. Indeed, the search for information about the impacts of nuclear energy helped define the emerging field of ecology. In this document—the summary of a report presented to the National Academy of Sciences—two prominent oceanographers draw some initial conclusions about whether or not nuclear wastes should be dumped at sea, a common practice in the early years of the Cold War by all countries with nuclear capabilities. How do Revelle and Schaefer propose to find answers to this question? What else do they see as necessary? What do they see as the role of science in answering questions about the viability of commercial nuclear power and the development of nuclear weapons? What do they hope to learn from this line of inquiry?

. . . .

The potential requirement for disposal of atomic wastes in the sea is sufficient reason for pursuit of these investigations. However, mankind

---

Excerpted from Roger Revelle and Milner B. Schaefer, "General Considerations Concerning the Ocean as a Receptacle for Artificially Radioactive Materials," in *The Effects of Atomic Radiation on Oceanography and Fisheries* (Washington, D.C.: National Academy of Sciences—National Research Council, 1957), 23–24.

will derive additional, and perhaps even greater, benefits in other ways. For example, the flux of materials through the various trophic levels of the biosphere is the fundamental process underlying the harvest of the sea fisheries. This process must be studied to provide part of the basis for atomic waste disposal, but its elucidation will also provide much of the scientific base for the optimum exploitation and conservation of the seas' living resources by man.

CONCLUSIONS AND RECOMMENDATIONS

. . . 1. Tests of atomic weapons can be carried out over or in the sea in selected localities without serious loss to fisheries if the planning and execution of the tests are based on adequate knowledge of the biological regime. The same thing is true of experimental introduction of fission products into the sea for scientific and engineering purposes.

2. Within the foreseeable future the problem of disposal of atomic wastes from nuclear fission power plants will greatly overshadow the present problems posed by the dispersal of radioactive materials from weapons tests. It may be convenient and perhaps necessary to dispose of some of these industrial wastes in the oceans. Sufficient knowledge is not now available to predict the effects of such disposal on man's use of other resources of the sea.

3. We are confident that the necessary knowledge can be obtained through an adequate and long-range program of research on the physics, chemistry, and geology of the sea and on the biology of marine organisms. Such a program would involve both field and laboratory experiments with radioactive material as well as the use of other techniques for oceanographic research. Although some research is already under way, the level of effort is too low. Far more important, much of the present research is too short-range in character, directed towards *ad hoc* solutions of immediate engineering problems, and as a result produces limited knowledge rather than the broad understanding upon which lasting solutions can be based.

4. We recommend that in future weapons tests there should be a serious effort to obtain the maximum of purely scientific information about the ocean, the atmosphere, and marine organisms. This requires, in our opinion, the following steps: (1) In the planning stage committees

of disinterested scientists should be consulted and their recommendations followed; (2) funds should be made available for scientific studies unrelated to the character of the weapons themselves; (3) the recommended scientific program should be supported and carried out independently of the military program rather than on a "not to interfere" basis.

5. Ignorance and emotionalism characterize much of the discussion of the effects of large amounts of radioactivity on the oceans and the fisheries. Our present knowledge should be sufficient to dispel much of the overconfidence on the one hand and the fear on the other that have characterized discussion both within the Government and among the general public. In our opinion, benefits would result from a considerable relaxation of secrecy in a serious attempt to spread knowledge and understanding throughout the population.

6. Sea disposal of radioactive waste materials, if carried out in a limited, experimental, controlled fashion, can provide some of the information required to evaluate the possibilities of, and limitations on, this method of disposal. Very careful regulation and evaluation of such operations will, however, be required. We, therefore, recommend that a national agency, with adequate authority, financial support, and technical staff, regulate and maintain records of such disposal, and that continuing scientific and engineering studies be made of the resulting effects in the sea.

7. We recommend that a National Academy of Sciences—National Research Council committee on atomic radiation in relation to oceanography and fisheries be established on a continuing basis to collect and evaluate information and to plan and coordinate scientific research.

8. Studies of the ocean and the atmosphere are more costly in time than in money, and time is already late to begin certain important studies. The problems involved cannot be attacked quickly or even, in many cases, directly. The pollution problems of the past and present, though serious, are not irremediable. The atomic waste problem, if allowed to get out of hand, might result in a profound, irrecoverable loss. We, therefore, plead with all urgency for immediate intensification and redirection of scientific effort on a world-wide basis towards building the structure of understanding that will be necessary in the future. This structure cannot be completed in a few years; decades of effort will be

necessary and mankind will be fortunate if the required knowledge is available at the time when the practical engineering problems have to be faced.

9. The world-girdling oceans cannot be separated into isolated parts. What happens at any one point in the sea ultimately affects the waters everywhere. Moreover, the oceans are international. No man and no nation can claim the exclusive ownership of the resources of the sea. The problem of the disposal of radioactive wastes, with its potential hazard to human use of marine resources, is thus an international one. In certain countries with small land areas and large populations, marine disposal of fission products may be essential to the economic development of atomic energy. We, therefore, recommend: (1) that cognizant international agencies formulate as soon as possible conventions for the safe disposal of atomic wastes at sea, based on existing scientific knowledge; (2) that the nations be urged to collaborate in studies of the oceans and their contained organisms, with the objective of developing comparatively safe means of oceanic disposal of the very large quantities of radioactive wastes that may be expected in the future.

10. Because of the increasing radioactive contamination of the sea and the atmosphere, many of the necessary experiments will not be possible after another ten or twenty years. The recommended international scientific effort should be developed on an urgent basis.

ATOMIC ENERGY COMMISSION

ATOMIC TESTS IN NEVADA, 1957

In the 1950s the Atomic Energy Commission (AEC) conducted one hundred aboveground nuclear tests at the Nevada Test Site in southern Nevada. The resulting mushroom clouds could be seen from over a hundred miles away. In 1957 the AEC distributed the pamphlet excerpted below to the people who lived in the vicinity. How did AEC officials justify the tests to the surrounding communities? How did they explain concerns about radioactivity? The drawing of the cowboy illustrated the pamphlet; why did the AEC try to evoke the mythology of the American West? What kind of messages do the cowboy and the mushroom cloud convey? What factors might have conditioned the members of the "downwind" communities to accept—or challenge—AEC claims?

The Nevada Test Site of the U.S. Atomic Energy Commission is used periodically for experiments or tests involving nuclear detonations of relatively low yield (explosive energy).

Forty-five nuclear fission weapons, weapon prototypes, and experimental devices were fired at the Nevada Test Site from January 1951 to June 1955. They ranged in yield from less than 1 kiloton up to considerably less than 100 kilotons. (A kiloton is equivalent to 1,000 tons of TNT.)

Despite their relatively low yield, Nevada tests have clearly demonstrated their value to all national atomic weapons programs. They have

---

Excerpted from United States Atomic Energy Commission, *Atomic Tests in Nevada* (Washington, D.C.: Atomic Energy Commission, 1957), 1–5.

made important contributions to the development of a whole family of weapons, including ones for defense against attack. Because of them our Armed Forces are stronger and our Civil Defense better prepared.

Each test fired in Nevada is justified, before it is scheduled, as to national need for the data sought. Each Nevada test has successfully added to scientific knowledge needed for development and use of atomic weapons, and needed to strengthen our defense against enemy weapons. Most tests have been used additionally for basic research, such as biological studies, which could be conducted only in the presence of a full scale nuclear detonation.

Conducting low-yield tests in Nevada, instead of in the distant Pacific, also has resulted in major savings in time, manpower, and money. The saving of time is particularly important because of its contribution to the Nation's defense capability.

PROTECTION OF THE PUBLIC

You people who live near Nevada Test Site are in a very real sense active participants in the Nation's atomic test program. You have been close observers of tests which have contributed greatly to building the defenses of our country and of the free world. Nevada tests have helped

us make great progress in a few years, and have been a vital factor in maintaining the peace of the world.

Some of you have been inconvenienced by our test operations. Nevertheless, you have accepted them without fuss and without alarm. Your cooperation has helped achieve an unusual record of safety.

To our knowledge no one outside the test site has been hurt in six years of testing. Only one person, a test participant, has been injured seriously as a result of the 45 detonations. His was an eye injury from the flash of light received at a point relatively near ground zero inside the test site. Experience has proved the adequacy of the safeguards which govern Nevada test operations.

POTENTIAL EXPOSURE IS LOW

Any atomic detonation, even though small enough to be fired in Nevada, involves powerful forces. The effects of a detonation include flash, blast, and radioactive fallout. Your potential exposure to these effects will be low, and it can be reduced still further by your continued cooperation.

The low level of public exposure has been made possible by very close attention to a variety of on-site and off-site procedures.

Public protection began with selection of the site. Nevada Test Site was selected only after extensive studies of other possible locations. The testing site covers an area of more than 600 square miles with an adjoining U.S. Air Force gunnery range of 4,000 square miles. The controlled areas are surrounded by wide expanses of sparsely populated land, providing optimum conditions for maintenance of safety.

EVERY TEST IS EVALUATED

Every test detonation in Nevada is carefully evaluated as to your safety before it is included in a schedule. Every phase of the operation is likewise studied from the safety viewpoint.

An advisory panel of experts in biology and medicine, blast, fallout, and meteorology is an integral part of the Nevada Test Organization. Before each nuclear detonation, a series of meetings is held at which this panel carefully weighs the question of firing with respect to assurance of your safety under the conditions then existing.

NATIONAL COMMITTEE FOR A SANE NUCLEAR POLICY

"WE ARE FACING A DANGER UNLIKE ANY DANGER THAT HAS EVER EXISTED," 1957

In 1957 the National Committee for a Sane Nuclear Policy, known as SANE, became the first grassroots group to stake out a strong public stance against the arms race. SANE announced its presence by taking out full-page advertisements in newspapers around the country—the text of this initial advertisement is included here. The group attracted a number of celebrities and high-profile members, especially from Hollywood and the entertainment industry. On what basis does SANE challenge the arms race? What are the consequences of pursuing a nuclear strategy? How does SANE use terms such as *progress*, *security*, and *natural rights*—and how are these terms used differently from the way that they might have been used to support the nuclear consensus? What role does the environment play in defining these terms?

• •

A deep uneasiness exists inside Americans as we look out on the world. It is not that we have suddenly become unsure of ourselves in a world in which the Soviet Union has dramatically laid claim to scientific supremacy.

Nor that the same propulsion device that can send a man-made satellite into outer space can send a missile carrying a hydrogen bomb across the ocean in eighteen minutes. . . .

Excerpted from an advertisement in the *New York Times*, November 15, 1957. SANE, Inc., Records, Swarthmore College Peace Collection. Courtesy of the Swarthmore College Peace Collection, Swarthmore, Penn.

The uneasiness that exists inside Americans has to do with the fact that we are not living up to our moral capacity in the world.

We have been living half a life. We have been developing our appetites, but we have been starving our purposes. We have been concerned with bigger incomes, bigger television screens, and bigger cars—but not with the big ideas on which our lives and freedoms depend.

We are facing a danger unlike any danger that has ever existed. In our possession and in the possession of the Russians are more than enough nuclear explosives to put an end to the life of man on earth.

Our uneasiness is the result of the fact that our approach to the danger is unequal to the danger. Our response to the challenge of today's world seems out of joint. The slogans and arguments that belong to the world of competitive national sovereignties—a world of plot and counter-plot—no longer fit the world of today or tomorrow.

Just in front of us opens a grand human adventure into outer space. But within us and all around us is the need to make this world whole before we set out for other ones. We can earn the right to explore other planets only as we make this one safe and fit for human habitation.

The sovereignty of the human community comes before all others—before the sovereignty of groups, tribes or nations. In that community, man has natural rights. He has the right to live and to grow, to breathe unpoisoned air, to work on uncontaminated soil. He has the right to his sacred nature.

If what nations are doing has the effect of destroying these natural rights, whether by upsetting the delicate balances on which life depends, or fouling the air, or devitalizing the land, or tampering with the genetic integrity of man himself; then it becomes necessary for people to restrain and tame the nations.

Indeed, the test of a nation's right to survive today is measured not by the size of its bombs or the range of its missiles, but by the size and range of its concern for the human community as a whole.

There can be no true security for America unless we can exert leadership in these terms, unless we become advocates of a grand design that is directed to the large cause of human destiny.

There can be no true security for America unless we can establish and keep vital connections with the world's people, unless there is some moral grandeur to our purposes, unless what we do is directed to the cause of human life and the free man.

There is much that America has said to the world. But the world is still waiting for us to say and do the things that will in deed and in truth represent our greatest strength.

What are these things?

FIRST, AS IT CONCERNS THE PEACE, AMERICA CAN SAY: That we pledge ourselves to the cause of peace with justice on earth, and that there is no sacrifice that we are not prepared to make, nothing we will not do to create such a just peace for all peoples;

That we are prepared to support the concept of a United Nations with adequate authority under law to prevent aggression, adequate authority to compel and enforce disarmament, adequate authority to settle disputes among nations according to principles of justice.

NEXT, AS IT CONCERNS NUCLEAR WEAPONS, AMERICA CAN SAY: That the earth is too small for intercontinental ballistic missiles and nuclear bombs, and that the first order of business for the world is to bring both under control;

That the development of satellites or rocket stations and the exploration of outer space must be carried on in the interests of the entire human community through a pooling of world science.

AS IT CONCERNS NUCLEAR WEAPONS, AMERICA CAN SAY: That because of the grave unanswered questions with respect to nuclear test explosions—especially as it concerns the contamination of air and water and food, and the injury to man himself—we are calling upon all nations to suspend such explosions at once;

That while the abolition of testing will not by itself solve the problem of peace or the problem of armaments, it enables the world to eliminate immediately at least one real and specific danger. Also, that the abolition of testing gives us a place to begin on the larger question of armaments control, for the problems in monitoring such tests are relatively uncomplicated.

AS IT CONCERNS OUR CONNECTIONS TO THE REST OF MANKIND, AMERICA CAN SAY: That none of the differences separating the governments of the world are as important as the membership of all peoples in the human family;

That the big challenge of the age is to develop the concept of a higher loyalty—loyalty by man to the human community;

That the greatest era of human history on earth is within reach of all mankind, that there is no area that cannot be made fertile or habitable, no disease that cannot be fought, no scarcity that cannot be conquered;

That all that is required for this is to re-direct our energies, re-discover our moral strength, re-define our purposes.

ATOMIC ENERGY COMMISSION

ATOMS FOR PEACE U.S.A, 1958

A central goal of the Atoms for Peace program involved public relations—convincing Americans that they had more to gain from the peaceful uses of nuclear energy than they had to lose from the fear of nuclear weapons. This document comes from a large-format, heavily illustrated book distributed by the AEC for just this purpose. The authors discuss the many ways that nuclear technologies reinforced consumer culture by making stronger plastics, healthier food, and other consumer amenities. How do AEC officials want the general public to view the potential of nuclear energy? Do they describe a radically new force, or something more commonplace? How might people expect nuclear energy to transform their everyday lives? What vision of progress does the AEC lay out in this document, and how is this vision tied to nuclear technology?

• •

Considering that there have been few scientific advances as revolutionary as the breakthrough from chemical to nuclear energy, it is not surprising that there should be more in the way of return than a new means—however important—of raising steam and generating electricity. The fact, of course, is that the first returns from the peaceful atom have had nothing to do with heat or power. They have come instead from the use of isotopes and radiation.

Excerpted from U.S. Atomic Energy Commission, *Atoms for Peace U.S.A. 1958.* A pictorial survey prepared by Arthur D. Little, Inc., for the United States Atomic Energy Commission (New York: Arthur D. Little, 1958), 111–16.

The use of isotopes is not new. . . . What *is* new is the scale on which they are being used. . . . For, while certain radioisotopes continue to be produced in cyclotrons—a method used since the early Thirties—the nuclear reactor has made them available in great variety and at low cost, putting on a routine assembly-line basis what had previously been a fairly esoteric laboratory activity. . . .

This point is impressively seen in the current statistics of isotope usage. One that comes first to mind is the number of groups and individuals working with radioisotopes. Current U.S. listings show some 1,500 industrial concerns, 2,000 medical institutions and independent physicians, and 250 colleges and universities—numbers that represent increases of from 75 to 150% in three years. Another is the curie volume of radioisotopes shipped out annually. Here one finds a five-fold increase (from 30,000 to 160,000 curies) over the past three years. Still another is the estimate of annual savings realized through the use of radioisotopes in U.S. industry. The Atomic Energy Commission's most recent figure: $500 million. . . .

But what is most impressive about isotopes cannot be expressed in numbers. It is the broad spectrum and remarkable variety of the problems they are helping men solve. This is reflected in the following excerpts from the alphabetical index of a recent ten-year survey of the industrial use of radioisotopes published by the Atomic Industrial Forum:

Air Conditioning
—leak detection
Bearings
—wear studies
Cigarettes
—density control
Detergents
—evaluation tests
Electric Light Bulbs
—envelope thickness control
Neon Lamps
—ionization source
Oil
—lubrication studies
Petroleum
—well logging
Quartz
—radiation effects
Semiconductors
—temperature stability studies

Fertilizer Manufacture
  —*process control*
Glass
  —*corrosion studies*
Hydrocarbons
  —*catalyst research*
Ink
  —*printing control*
Kinetics
  —*study of reaction rates*
Locomotives
  —*radiographic inspection*
Minerals
  —*study of flotation processes*
Tanks
  —*liquid-level indicators*
Ultrasonics
  —*evaluation of cleaning action*
Valves
  —*efficiency studies*
Waxes
  —*testing of*
X-rays
  —*portable source*
Yeast
  —*metabolism studies*
Zinc
  —*activation analysis of alloys*

This list could be multiplied many times, but it would consist of variations on four principal themes: tracing, measurement and control, radiography, and teletherapy.

The use of radioisotopes as tracers has added a new dimension to research in the physical and life sciences and opened up new avenues of medical diagnosis and therapy. It has streamlined countless industrial operations, ranging—in the petroleum industry, for example—from logging oil wells, to tracing catalyst flow in refining, to monitoring the flow of refinery products through pipelines. Used in such industrial measurement and control devices as thickness, density and liquid-level gages, radioisotopes have reduced wastage and improved uniformity in various basic manufacturing industries. For example, beta gages are used to control the thickness of 90% of the automobile tire fabric and 80% of the tin plate produced in the United States. Nine out of ten of the country's major paper manufacturers use these gages to control the uniformity of their products, and indeed the paper on which these words are printed has been so monitored. . . .

There has arrived, in short, what Willard F. Libby, originator of the technique of dating archeological artifacts by carbon-14 measurement and since 1954 a member of the United States Atomic Energy Commission, has called "the golden age of isotopes."

A separate field of activity is research on massive radiation applications on the chemical, food, and drug industries. Here fundamental changes in properties are involved. Here the possibilities are enormous, but progress does not come so easily.

One of the first problem areas to be attacked in this field was the preservation of food by irradiation. It had long been known that ionizing radiations have the ability to destroy microorganisms and enzymes—both major causes of food spoilage. It was originally hoped that by high-dosage gamma irradiation food products could be "cold-sterilized" to a degree that would permit unrefrigerated storage for an indefinite period. This has indeed been done, and with reproducible results; however, it has frequently been accompanied by off-flavor effects and other undesirable side reactions. While progress is being made toward an understanding of these unwanted effects, the emphasis has shifted to low-dosage treatment, often in combination with conventional preservation techniques, with the more limited objective of increasing shelf life for finite periods. This approach, analogous to pasteurization, has been found very effective and is today the principal basis for a widespread food-preservation research effort involving something like one hundred university and industrial laboratories in the United States alone. . . .

Another is the work that has been done on the use of radiation to vulcanize rubber. . . . "Radiovulcanization" has been achieved by simple exposure of uncured rubber at room temperature to the radiation given off by spent fuel elements from a nuclear reactor. . . . Automobile tires treated in this fashion have shown improved resistance to wear and abrasion; however, in this instance the amount of radiation involved and its current high cost appear to rule out commercial application for the present. . . .

At the present early stage of radiation chemistry research it is impossible to predict which of the many possible avenues of massive radiation application will be the most fruitful, let alone the answer to the question of which sources will be in greatest demand. What is perhaps most significant at this time is the amount of research effort being put into this field. The Atomic Energy Commission is sponsoring an intensive program, including basic research on the mechanism of radiation effects on chemical reactions, applied research on specific applications of industrial interest, and work on methods of reducing the cost of fission-product sources. And, as was brought out in the earlier review of the U.S. research and development network, most of the country's larger oil and chemical companies are hard at work on radiation studies, several of them in new and extensively equipped radiochemical laboratories.

Returning to the broad field of massive radiation uses, it is clear that the problems are generally difficult. But as an industry spokesman said recently after presenting a sober assessment of the progress thus far:

> "It is most important to recall that literally thousands of competent research workers are spending their time in the area and it is rare indeed that such a concentrated effort of technical skills is denied results."

BARRY COMMONER

"THE FALLOUT PROBLEM," 1958

In the late 1950s biologist Barry Commoner emerged as a leading spokesman for the antinuclear movement. Spurred by his early concern about the ecological impacts of radioactivity, he became one of the nation's most well-known environmental activists. He helped found the Greater St. Louis Citizens' Committee for Nuclear Information, which raised awareness and concern about the impacts of nuclear testing. In this article Commoner reflects on the 1957 congressional hearings on the biological impacts of fallout and considers how scientists, politicians, and citizens might use this information to create sound public policy. What does Commoner see as the role of science in policy-making, and what does he see as the impact of scientific uncertainty? Who does he think should be making decisions about such topics as nuclear testing—and based on what kind of information?

• •

The fallout problem results from the decision on the part of three governments to carry out a particular type of military activity: test explosions of nuclear weapons. It is reasonable to expect these governments to determine that nuclear tests shall not cause inadmissible hazards to human life. This responsibility requires: (i) determination of the need for, and the advantages to be derived from, the nuclear operations; (ii) estimation of the extent and character of the hazards; and (iii) a judgment of the relative weights of the advantages and hazards.

Excerpted from Barry Commoner, "The Fallout Problem," *Science* 27, no. 3305 (May 2, 1958): 1023–26. Reprinted with permission from AAAS.

In an orderly state of affairs we expect scientists to produce an evaluation of possible dangers sufficiently clear and sufficiently close to being unanimous to provide a workable basis for decision. We expect the makers of policy to reach a conclusion which represents a balanced evaluation of needs and hazards. We expect the public to be sufficiently informed about the needs and possibly harmful consequences to understand and support this judgment.

This is the ideal. What is the reality?

That governments find advantage in conducting test nuclear explosions may as well be taken here as a fact of political life. It is not our purpose at this time to debate the validity of this need. . . .

Although the Congressional hearings did not consider how the advantages of nuclear testing might be weighed against the estimated hazards, the evidence heard gives us a picture of the size of the problem. Anyone who attempts to determine whether or not the biological hazards of world-wide fallout can be justified by necessity must somehow weigh a number of human lives against deliberate action to achieve a desired military or political advantage. Such decisions have been made before—for example, by military commanders—but never in the history of humanity has such a judgment involved literally every individual now living and expected for some generations to live on the earth.

It is not clear who is expected to make this decision and thereby assume, in an unprecedented degree, the grave moral burden carried by those who must judge the social worth of human life. Should this judgment be made by experts with special competence? If so, where should their expertness lie? In nuclear physics, radiochemistry, biology, medicine, sociology, military strategy? On the other hand, should a responsibility of this weight be reserved to elected officials, in order to ensure that the decisions will reflect the ethical views of our society? At the moment, we seem to have no stated policy in this matter.

Finally, the present situation is also unsatisfactory with regard to the state of public knowledge on the fallout problem. This conclusion could be documented in many ways, but perhaps the most objective and significant view is that contained in the Joint Committee's summary-analysis: "Information on fallout has evidently not reached the public in adequate or understandable ways."

Besides being poorly informed, the public has been confused by disagreements among scientists regarding the biological danger of present and anticipated radiation levels from fallout. The public is accustomed to associating science with truth and is dismayed that scientists appear to find the truth about fallout so elusive. . . .

Why do scientists disagree in their estimates of the biological hazard of world-wide fallout?

The scientific problem is extraordinarily difficult and complex. Its solution requires an understanding of vast interactions among masses of air, water, and soil and innumerable varieties of plants, animals, and men. Compared with our knowledge of other agencies that affect life, such as light and heat, our knowledge of ionizing radiation is recent. In the scant 60 years since the discovery of radiation, there has not been time enough for biologists satisfactorily to explore its effect on life. Strontium-90, the chief source of fallout radiation, is an element only recently made by man; there has been little time to study it in the laboratory or to analyze the consequences of its intrusion into nature. Finally, the major hazards of radiation—cancer and genetic mutation—are perhaps the most difficult unsolved problems of modern biology. Until the basic causes of these processes become more clear, the effects of radiation will be but poorly understood.

In this situation the available facts are often not sufficient to support or contradict conclusively a given explanatory idea; therefore, opposing theories will for the time flourish together. This accounts for some of the disagreement among scientists' estimates of the probable biological hazard of fallout radiation.

In part, our present troubles derive from the unequal pace of the development of physics and biology. We understand nuclear energy well enough to explode great quantities of radioactive materials into the atmosphere. But our present knowledge of biology and its attendant sciences is not adequate for contending with the difficulties that follow when the radioactive dust settles back to earth. . . .

The inadequacies which decision makers now find in basic and operational information about fallout hazards are, then, in part due to the lack of detailed, integrated, continuing data published in a form capable of enlisting the interest of the entire scientific community in this pervasive problem. We need to recall that the development of a scientific truth is a direct outcome of the degree of communication which normally exists in science. As individuals, scientists are no less fallible than any other reasonably cautious people. What we call a scientific truth emerges from investigators' insistence on free publication of their own observations. This permits the rest of the scientific community to check the data and evaluate the interpretations, so that eventually a commonly held body of facts and ideas comes into being. Any failure to communicate information to the entire scientific community hampers the attainment of a common understanding.

In sum, the fallout question has not yet become an integral part of the freely flowing stream of information which is the vehicle of scientific progress. The remedy is apparent and, for scientists, traditional: more and better publication in readily available journals. Can we not establish a systematic method of continuously reporting integrated information on world-wide levels of fallout radioactivity? . . .

What we need now is to marshal the full assemblage of facts about fallout, their meaning and uncertainties, and report them to the widest possible audience. This is not an easy task. It is much simpler to publicize conclusions alone, and have them accepted not because their factual origin is fully understood but because they carry the authority associated with science.

It seems to me that we dare not take this easy way out. Unless the public has sufficient information to provide a reasonable basis for independent judgment, the moral burden for the future effects of nuclear testing will rest on some smaller group. And no such group alone has the wisdom to make the correct choice or the strength to sustain it. Unless the public is made aware of the gaps and the uncertainties in our present knowledge about fallout, we cannot expect it to support the expensive research needed to minimize them. Without public understanding and support, no government policy can long endure.

Here then is our challenge. Can we, as scientists, with the help of our professional organizations, find a way to inform the public about these great issues? The raw material for such an educational campaign is available in the voluminous report of the Congressional hearings. We can distill from this material the essential facts and ideas and bring them to the people through the media of public communication: radio and television, newspaper articles, and widely distributed pamphlets.

In sum, here are the tasks which the fallout problem imposes upon us. Research into the hazards of fallout radiation needs to be more fully and widely published so that the scientific community will be constantly aware of the changes which world-wide radiation is making in the life of the planet and its inhabitants. This knowledge must be at the ready command of every scientist, so that we can all participate in the broad educational campaign that must be put into effect to bring this knowledge to the public. If we succeed in this we will have met our major duty, for a public informed on this issue is the only true source of the moral wisdom that must determine our nation's policy on the testing—and the belligerent use—of nuclear weapons.

There is a full circle of relationships which connects science and society. The advance of science has thrust grave social issues upon us. And, in turn, social morality will determine whether the enormous natural forces that we now control will be used for destruction—or reserved for the creative purposes that alone give meaning to the pursuit of knowledge.

EDWARD TELLER

"THE PLOWSHARE PROGRAM," 1959

Perhaps the most fantastical plans for the peaceful uses of nuclear energy involved the government program known as Project Plowshare. "Geographical engineering"—as Edward Teller, the project's leader, called it—involved using controlled underground nuclear and thermonuclear explosions to construct canals, deepen harbors, and release natural gas, as well as to complete other large-scale projects. Teller worked as a physicist for the Manhattan Project during World War II and served as the greatest scientific champion for a strong nuclear program, as evidenced by his support of Project Plowshare and the development of thermonuclear weapons. Project Plowshare exploded nearly thirty underground bombs around the United States, but the program proved controversial because of fears of environmental and human health risks and cost overruns. Public opposition led to the cancellation of the program in 1977.

What kind of attitudes about nature do Teller's plans for geographical engineering reveal, and how are these attitudes related to ideas about science and progress? What does he say is necessary to bring Project Plowshare to fruition? How do his attitudes about science—and the pursuit of knowledge—compare to those advanced by Roger Revelle, Milner Schaefer, and Barry Commoner?

Excerpted from Edward Teller, "The Plowshare Program," in University of California Lawrence Livermore Laboratories and San Francisco Operations Office, U.S. Atomic Energy Commission, *Proceedings of the Second Plowshare Symposium* (Washington, D.C.: U.S. Atomic Energy Commission, 1959).

Conventional explosives have been used in peacetime as much, perhaps more, than they have been used in wartime. They have been used in the main for two purposes. They have been used for transportation in the sense that obstacles have been removed in building roads, waterways and other similar things. And they have been used in mining. You will see that, according to our present ideas, these are also the two main uses of nuclear explosives, although you will hear about other classes of possible applications.

The reason that nuclear explosives have been so slow in entering our ideas for peaceful purposes is twofold. One reason is that nuclear explosives have sort of biggish effects with the result that one tends to think twice or perhaps even more than twice before one actually tries to undertake something. The other reason is that nuclear explosives produce radioactivity. This radioactivity in the immediate neighborhood of the nuclear explosive can, indeed, be quite dangerous; one has to take, and we are going to take . . . the most conscientious precautions that this radioactivity will do no harm. Of course, one should also realize that radioactivity in exceedingly small doses (a subject that often enters the public discussion) is really far from dangerous; one cannot even determine sometimes whether a danger exists.

I should say right away that many of the applications are underground explosions where the radioactivity is completely contained and where, provided one pays serious attention to the question of possible ground water contamination, the danger will be quite negligible. Furthermore, I should tell you that the radioactivity we are likely to release underground is very much smaller than the radioactivity that one buries when one tries to dispose of reactor products. The difficulties due to possible contamination merit serious and conscientious consideration. But they are by no means insurmountable. . . .

I have told you that two great difficulties connected with Plowshare are the fact that nuclear explosives are big and that they produce radioactivity. I would like to assure you that we are working very hard on these two topics. It stands to reason that the full effect of nuclear explosives can be exploited only if the biggest kind of nuclear explosions are included in our program and in the long run we hope to include them. At the same time it will add a very great deal to the flexibility and the scope of the program if smaller nuclear explosions can be carried out and particularly if they can be carried out cheaply. This is one of the aims of continued research into nuclear explosives and I only wish I could tell you more about this exceedingly interesting topic. Perhaps someday I shall be able, or somebody shall be able, to do so.

Second, I should like to tell you that the difficulty of radioactive contamination is also something which we can attack. Even at the present time we know how to make big nuclear explosions fairly clean. We are hard at work to make big and small nuclear explosions as clean as possible and, hopefully, we may make them completely clean. It will require invention, it will require work, but if we could get rid of radioactive contamination, the applicability of this new tool would be very, very greatly increased, and this is one of the most important aims toward which our experiments are, and must be, directed.

Now, as to the program itself, we need help. We have knowledge—good knowledge, reliable knowledge—about nuclear explosives: how to make them, how to handle them, how to apply them. We acquired some knowledge of the behavior of nuclear explosives in the air quite some time ago, and underground more recently. This knowledge comes in very handy, but we have to have more knowledge, particularly about underground nuclear explosions.

However, there are very, very big areas where we are in dire need of advice and cooperation; in fact, this program cannot prosper without a great number of people, earth scientists, chemists of every description, scientific people of every kind. If these people will participate, and I know many of you are in these various fields, then the program will prosper; otherwise it will not.

Let me sketch briefly the state of our prospects and ideas. First of all, we have the earth-moving projects—building harbors, building canals. To move earth, and to move earth cheaply, is something that we know we can do today. We do not know the best way to do it, we do not know in every detail how the radioactivity can be handled, and we can only hope that the radioactivity might be avoided altogether, as I have told you. . . . These are tasks which we can perform, which we hope to perform wisely with the help of a wide and thorough discussion both with respect to safety and with respect to the question of whether the projects are worthwhile.

Then, there are many projects which do not depend on the moving of earth but which depend on the breaking up of rock. These include possible regulation of underground flow of water, storage of water, and all manner of mining including oil production. These two words, water and oil, are magic. Oil is our most important fuel and the lifeblood of modern industry. It may be a strange reversal when nuclear energy will be used to make the main source of conventional energy cheaper and more plentiful. We seem to be putting some cart before some horse. . . .

There is a big area of, in my opinion, even more adventurous thought. We can produce heat and we can mine the heat. If we can make the capital investment of piping and other things serve for many shots in the same place, we can hope to do this in a really economical way and to have really cheap and abundant thermonuclear power. We can make isotopes of all descriptions underground; some of these isotopes will be very valuable, provided we can extract them in a cheap and effective manner.

Now I would like to mention to you two fields which, I believe, have been considered much too little. They have been considered, but the consideration has not been sufficient. One is the use of nuclear explosions for purely scientific purposes. A nuclear explosion underground is a source of geophysical disturbance, a smallish earthquake which can be used to find out more about the crust and interior of our earth. A nuclear explosion is a source of radiation and neutrons; we can use nuclear explosives to find out more about neutrons and nuclear physics in general. One can use a nuclear explosive as it was used unintentionally in the first big nuclear explosion in the Pacific: to capture many successive neutrons in the same nucleus and thereby get to a new species which then can be studied. One can use nuclear explosives for various strange purposes to find out the properties of the upper atmosphere and the earth's magnetic field. . . .

This group here, if you'll pardon me for saying so, seems to be similar to a bunch of starved and blind people stumbling around in an extremely rich garden. The food, I'm sure, is there to pick; the only question you have to decide is in which direction you want to extend your hand; then you have to rely on the best methods of touch, smell, and intelligence lest you bite into a bitter fruit. I am sure that the opportunity is here. I am sure that somewhere among you there is the idea, the initiative, the perseverance and the genius which will bring this Plowshare program to fruitation.

OFFICE OF CIVIL DEFENSE AND MOBILIZATION

FALLOUT MAPS, 1959

These maps appeared in a 1959 publication by the Office of Civil Defense and Mobilization titled "The Family Fallout Shelter." They predicted the likely patterns of fallout one hour and twenty-four hours after a nuclear attack. Maps like these changed the tenor of public discussions about fallout and nuclear war, causing people to doubt the survivability of war and to question the basic assumptions of the arms race. What kind of reactions might these maps have generated?

• •

"The Family Fallout Shelter," Office of Civil Defense and Mobilization (Washington, D.C.: U.S. Government Printing Office, 1959).

Figure 1.—Fallout areas at 1 hour after detonation

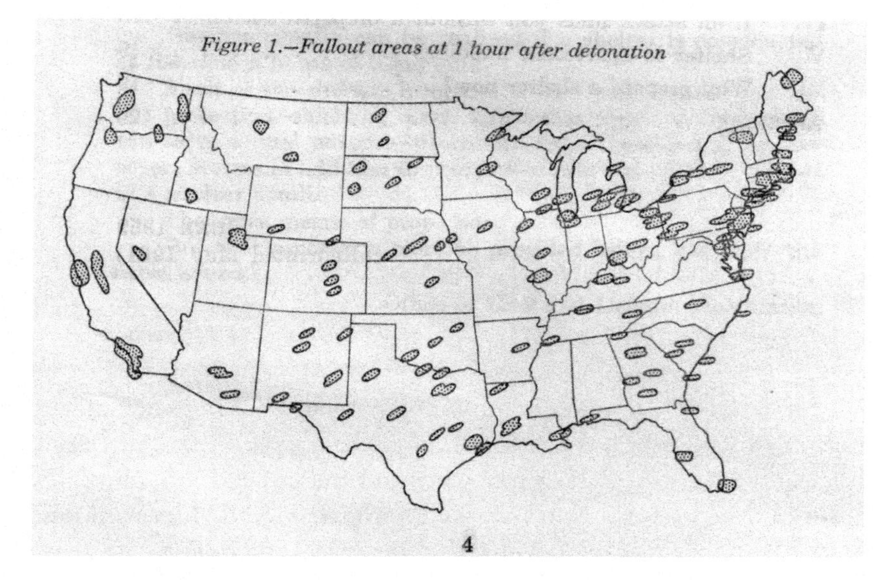

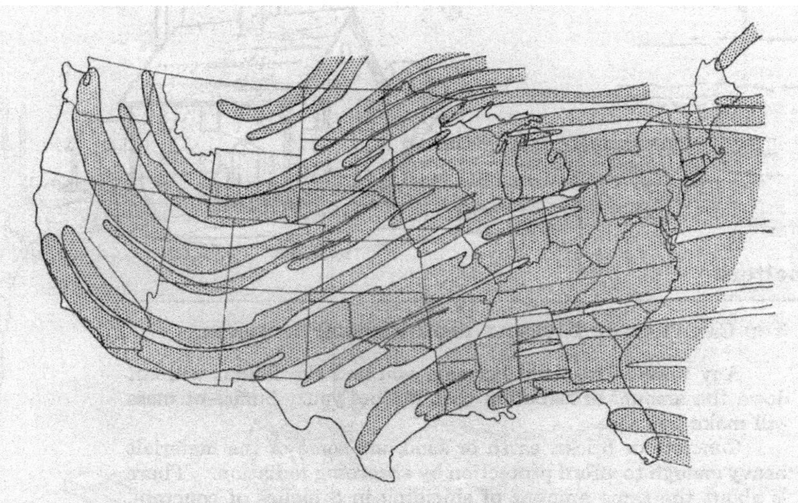

Figure 2.—Fallout areas at 24 hours after detonation

HERMAN KAHN AND H. H. MITCHELL

THE POSTATTACK ENVIRONMENT, 1961

In the early 1960s, with Cold War tensions reaching their peak, military strategist Herman Kahn and others urged that American nuclear options needed to move beyond the concept of "mutually assured destruction." Kahn worked with the RAND Corporation, an influential think tank devoted to geopolitical and military strategy. In his controversial 1960 book *On Thermonuclear War* (published by Princeton University Press), Kahn argued that although it might be unthinkable, a nuclear war was winnable. If the United States took appropriate preparatory measures, the basic economic, political, and social structures of the nation could survive even a thermonuclear war. This theory depended on speculation about the "postattack environment" and prompted significant research by RAND Corporation scientists and other investigators into the eventual ecological impacts of nuclear war. What kinds of postattack environmental issues do Kahn and his RAND Corporation colleague H. H. Mitchell take into account? What parallels exist between the way that these men view the postattack environment and other predictions of environmental destruction from non-nuclear causes?

Excerpted from Herman Kahn and H. H. Mitchell, Subcommittee of the House Committee on Operations, *Civil Defense—1961*, 87th Congress, 1st session, 1961, 170–73, 177–78, 331–33, 337.

[Herman Kahn:]

. . . Can civil defense be used to protect lives, protect property, or to facilitate recuperation after a war is over?

Rather surprisingly, I think most people have the impression that the answer to this question is also "No." Let me show you a chart that indicates some reasons for this point of view. (See fig. K-2.)

TABLE K-2. Tragic but Distinguishable Postwar States

DEAD	RECUPERATION (YEARS)
2,000,000	1
5,000,000	2
10,000,000	5
20,000,000	10
40,000,000	20
80,000,000	50
100,000,000	100

Will the survivors envy the dead?

This chart . . . is entitled "Tragic but Distinguishable Postwar States." I mean by this title that it should be possible for any person to distinguish between wars which result, say, in 20 million dead Americans and wars which result in 40 million dead Americans. I mean further that if this American has to choose between a war with 20 million dead Americans and one which results in 40 million dead Americans, other things being equal, he prefers a war in which there would be 20 million dead Americans.

It is practically impossible for people to make such a remark, to actually say, "I prefer a war which results in 20 million dead rather than one which results in 40 million dead." They want to say, "Why should I have a war at all?" They want to know, "Why do you tell me to choose between a war in which there will be 20 million dead Americans and a war in which there will be 40 million dead Americans? I do not want to choose." Because they do not want to choose, they choose by default. They ignore the problem. . . .

THE POSTATTACK ENVIRONMENT

There is one important question which is raised by the effects of modern weapons today, the question of survivors envying the dead. This is an important point; the most frightening possible point. This question is raised mainly because modern weapons have long-lasting effects.

. . . I will not surprise you when I say, perfectly soberly and reasonably, that if there is a war today the environment will be more hostile to human life for, say, 10,000 years.

Now, one can argue this statement; it might be wrong. However, the best scientific evidence indicates that it is correct, that the environment will be more hostile to human life for 10,000 years or so. I am thinking now of the longer lived, radioactivity due to carbon 14.

To many people this statement carries the implication that it is not worth living in that hostile environment. That is, of course, much too quick and shallow an opinion. . . . [W]hile the environment would be more hostile, it would not be so hostile as to preclude normal and happy lives for the survivors.

In other words, the survivors can rebuild, they can reconstruct and, in many cases, they would not notice the greater hostility of the environment. It would be a statistical effect which would be discernible in the mortality tables, but not by the average individual's personal observation.

The average individual would go through life running somewhat greater risks of various types of diseases and greater risks of having genetically deformed children, but when these risks are compared to the risks normally run today, they are not startlingly larger. The quality of life, would not necessarily have been changed dramatically. . . .

LONG-TERM RECUPERATION

The next problem is the long-term recuperation problem. Recuperation here has many facets: economic, social, political, psychological, and, in a subtle way, moral.

The only one which has been studied with any care is the economic. Here I think we can say with some confidence that . . . the economy will come back with amazing resilience; in other words, countries like the United States are extremely competent, once they get started, at producing capital and consumer goods. . . . but, depending upon the war, one would conjecture that we could rebuild the destroyed wealth in less than a generation, in all likelihood, in, say, 10 years after the kind of war that is usually envisaged.

. . . [In] terms of studies which have to be done and in terms of the most serious questions which remain unanswered, these social, psychological, political, and moral questions are currently the hard questions. Many feel they are the dominating questions.

However, it is my personal belief, speaking less as an expert than as a man who has read widely, that these problems have been grossly exaggerated.

Most people will not be psychologically deranged. One is not, for example, going to break up family relationships by a war. The family relationship is a very stable one.

One is not even going to obliterate the basic fact that people are Americans. By and large, they will be about as honest, hard-working, reliable, and responsible as they are today. While everybody's lives and thoughts will be affected by the war, the character structure of the survivors is unlikely to be changed in any startling fashion.

The political questions are more difficult. We live today in a very stable country. It is one of the few countries in the world in which the government does not worry about revolution and subversion as major problems, because we do not expect the Government to be subverted or overturned. However, such an event could occur as a result of a war. Even if we won the war, it is conceivable that we might no longer live in a democracy.

However, even though a war is a cataclysmic event, it seems to me a reasonable conjecture, particularly if preparations have been made, that our political democracy could survive most wars. But this statement has more faith in it than analysis. It is not a statement which I would try to maintain before a hostile and skeptical audience. I hope, however, that this and similar statements will soon be subjected to a more careful and deeper examination than I am capable of giving them. . . .

[H. H. Mitchell, M.D.:]

ECOLOGICAL PROBLEMS AND POSTWAR RECUPERATION

My purpose in appearing before the committee is to call attention to one of the most important but as yet unemphasized areas of civil defense; namely, assessing and solving ecological problems of the postattack environment. At present one can only try to pose questions properly and suggest areas for research because in its major aspects, this has been a strangely neglected field. . . .

Ecology may be defined as the study of the relationships of populations of organisms (plant and animal) to each other, along with the effects of the physical and chemical environment on these relationships. For our immediate purpose we are interested in how disturbances caused by a nuclear attack will affect man's ability to exist because of possible failures in the biological-environmental complex. It is analogous to the problem in the industrial economy associated with natural resources, stockpiling, and bottlenecks.

The interactions of populations of living organisms and their relationships to the environment make up a dynamic system, with living and nonliving substances being moved about in what is known as an ecosystem. This is the fundamental unit of ecology, and it is within this unit that we will be looking for problems relating to postattack survival and recuperation. Nuclear war might conceivably lead to complete sterilization of life in a particular area because of fire and radioactivity. Or there could be a selective removal of one or more essential biotic elements which could have significant sequential effects (e.g., removal of higher plants leading to floods and erosion and followed by decreased agricultural output later).

It is important to realize that we in the United States are in some sort of rough equilibrium with most of our ecosystems. There is a flow of food, fibers, and other material into the economy of man and there are also various levels of control over harmful aspects such as disease and infestation. Prevention of animal and plant disease involves ecological principles. Insect infestations are also a major concern. Disturbance of established relations could lead to serious unexpected consequences for man in the postattack environment. Even in peacetime, ecological dislocations can be serious. An interesting example involving insects is the population oscillation of the locust in the Asiatic Middle East. The locust lives in desert or semi-arid country and in most years is non-migratory and eats no crops. At intervals for reasons not completely understood, the population density greatly increases. The locust actually undergoes anatomical changes and starts to emigrate into cultivated lands, eating all crops in its path. This is the type of phenomenon which could occur in the disturbed conditions of our postwar environment, and the risk of insect infestation, its consequences, and amelioration should be studied in detail.

The main direct effects of nuclear weapons on various ecosystems will be from fire and fallout radiation. Fire, of course, will have a direct effect by burning forests, grasslands, wildlife and livestock. The indirect consequences of this must also be examined. Radiation will affect various species of plant and animal life directly, and different results may be expected at various levels of radiation. Another effect of radiation is the passage of isotopes through the food chain to final deposition in man. A great deal of attention has been given to this effect because of the interest in fallout from tests and its hazards to man. However, we want to examine this problem from a broader ecological point of view and assess such radiation hazards as the possibility of the soil becoming sterilized through the destruction of decomposers (bacterial and fungi), the inability of crops to grow, or the upsetting of population balance between two or more life forms because of differential radiosensitivity. . . .

What seems to come out of all this is that an inventory and research effort in the "biological economy" sector should be instituted at a level of intensity comparable to that going on in the industrial economy (from the postattack recuperation point of view). The resources of the agricultural bureaus of Federal and State Governments, as well as academic departments in universities and agricultural schools, undoubtedly contain information and personnel which could be brought to bear on the problems of concern to us. The Department of the Interior, the Forestry Service and the Army Corps of Engineers are still other agencies whose knowledge and skill should be utilized for study of the problems involved.

MARGARET MEAD

"ARE SHELTERS THE ANSWER?" 1961

The election of John F. Kennedy in 1960 sparked a renewed anxiety about the prospects of nuclear warfare. Kennedy's aggressive Cold War rhetoric and his requests for a dramatically increased military budget and a nationwide civil defense program prompted widespread public discussion about the construction of fallout shelters. This discussion reflected the development of thermonuclear bombs and the recognition that surviving the fallout after a nuclear attack would prove to be as great a challenge as surviving the attack itself. Public debate also concerned the ethics of fallout shelters: who should be allowed in, and to what extent should those inside the shelter protect themselves from those left outside?

Anthropologist Margaret Mead—one of the nation's foremost intellectual voices—addressed some of these ethical questions in the article below. She offers a strikingly different vision of citizenship and civil society than the one that emerged from the civil defense initiative of the previous decade. How does she assess the shelter scare? Does she root her arguments in biology or culture? What anxieties does this article reveal about nuclear weapons, the Cold War, and American society in general?

In the cold war of nerves, the state of morale in the opposing countries is an essential ingredient of defense, part of the huge paraphernalia of

---

Excerpted from Margaret Mead, "Are Shelters the Answer?" *New York Times Magazine* (November 26, 1961): 29, 124–26. Reprinted by permission of Mary Catherine Bateson.

deterrence each side has assembled. But given the differences between the American and the Soviet systems, each country necessarily goes about building civilian morale differently. In contrast to the Soviet Union, we have announced publicly that civilian defense is part of our program of deterrence.

From one point of view, the assurance that Americans have the will and the intention to survive any kind of attack should introduce one more beneficent pause into the Soviet Union's calculations. There are, however, some questions. Can the civilian defense program we are now developing really produce the steadfastness, the resolute determination neither to surrender nor to allow ourselves to be panicked into provocation or attack? Does it convey to those who might attack us that we are—or are not—expecting to be attacked? Is the program, in fact, a deterrent or a provocation?

This past summer Western Europe resonated with horror to reports that some Americans were planning to build individual bomb shelters in which each man would defend his own family, by force of arms if necessary, against his less provident neighbors. Many Europeans who had survived the agonies of warfare against civilians, sharing space and air with all who needed shelter, took these stories as one more example of Americans' inexplicable affinity for violence. In their eyes, Americans had accepted the idea of war—a war that would destroy the rest of mankind—for the sake of a set of principles that, if these stories were to be believed, would not stand up to a single bombing.

Set beside such terms as "overkill" and "megadeath," the picture of members of the richest and most technically advanced country in the world regressing to a level lower than that of any savages, to the level of trapped animals, made Europeans shudder. Perhaps the world did not have to wait for a nuclear war to bring about the physical dissolution of civilization: perhaps it was dissolving morally now.

New stories added new details to the picture. A clergyman sanctioned the right of a man to kill his neighbor in order to protect his own family—a right accorded to no member of a society which calls itself a society. A nationwide television program, depicting the fictional response to an alert, showed a frantic group of neighbors battering down in violent rage a family's shelter which, once destroyed, could protect no one.

What such stories suggested was that the Government, by giving its blessing to the construction of individual shelters, had abnegated its historical responsibility to defend the people of the United States and severed the ties that should bind together any men and their government in an emergency. Instead of fostering the shared responsibility that makes all men equal because all share the danger, the admonition to build individual shelters would make rich and poor, city dweller and suburban dweller, householder and tenant glaringly unequal. . . .

Yet even though we recognize the present shelter program and the American people's response to it to be symptomatic, we need not regard the current discussion as a total waste. The debate about shelters has caused an upsurge of genuine, realistic concern about nuclear war. Americans can refocus this concern into a greater understanding and acceptance of the responsibility carried by the nation which first invented the bomb and, so far, is the only one to have used it. In turn, this can become a mandate for a national effort to invent ways to protect the peoples of the world from a war which might end in the extinction of the human species.

The first step in the direction of a world rule of law is the recognition that peace no longer is an unobtainable ideal but a necessary condition of continued human existence. But to take even this step we must return to a calm and responsible frame of mind, a frame of mind in which we can face the long patient tasks ahead. This means finding a workable resolution of the present confused, panicky, incoherent discussion of the shelter problem itself.

To do this, what do we need? Most immediately, we need clarification of the facts. We need a clear, frank statement of the differences among blast effects, avoidable fall-out effects, and long-term death and mutation. We need as clear a formulation as possible of what the risks of an all-out nuclear war would be for ourselves and for every other people on earth, what the chances would be that, given the survival of some human beings, it would be possible to reorganize something resembling the free society we wish to protect.

This knowledge would help the American people to judge what it is we are trying to prevent by deterrence, how essential it is to find some safer solution. Judgment that is based on knowledge should strengthen our determination both to support our present policy of deterrence and to work unsparingly to end the spiraling arms race.

But the damage to our trust in our Government, our physical scientists, and ourselves cannot be undone simply by refocusing our concern onto the safety of the human race. Those who cannot protect their own people cannot undertake to protect others. . . .

The nation-state, advanced by warfare and given time to consolidate peaceful internal gains by periods of truce, was a great social invention on mankind's long road toward higher levels of political integration. Now we need a further social invention that will give us a way of extending our responsibility, based on our own nationhood, beyond our borders. We need a way of including not only our allies and the uncommitted nations but even our enemies, *while they are our enemies.*

No more than the father of a family guarding an individual shelter, or the little town arming itself against refugees from the city, can a single nation protect its people unless it also protects the rest of mankind, with or without their consent. This is a form of nationhood that is just on the edge of being invented. . . .

The more expansive we become in our willingness to take responsibility not only for our children but for all children, including the children of our enemies, the more we can mobilize our energies and the energies of the world to bring into being these new forms of nationhood.

WOMEN STRIKE FOR PEACE MILK CAMPAIGN, 1961

In November 1961 over fifty thousand women in more than sixty cities across the United States participated in a one-day protest known as Women Strike for Peace. In some communities women marched; in others they lobbied elected officials or published letters in local newspapers. After this day of action, a loosely organized movement persisted in trying to shape the national conversation about the arms race and the dangers of fallout. The women involved in this movement reacted in particular to concerns raised by the baby tooth survey conducted by Barry Commoner and the Greater St. Louis Citizens' Committee for Nuclear Information. The radioactive isotope Strontium-90 enters the human body through cow's milk. Strontium-90 mimics the chemical behavior of calcium and, when ingested, is deposited in bones and bone marrow. For those involved in the Women Strike for Peace movement, this raised concerns about the safety of the nation's milk supply and children's health. Why were the women concerned with milk? How did they use the threat posed to children's health to challenge the nuclear consensus? Which elements, in particular, of the nuclear consensus did they challenge? In what ways does their position compare to SANE and other peace movement voices?

Women Strike for Peace Milk Campaign flyers, Women Strike for Peace Papers, Box 1, Folder 4, Wisconsin Historical Society Archives, Madison, Wisconsin. Documents and image courtesy of the Wisconsin Historical Society.

Women

STRIKE FOR PEACE!

WE STRIKE AGAINST DEATH, DESOLATION AND DESTRUCTION.

WE STRIKE FOR LIFE, LIBERTY AND THE PURSUIT OF HAPPINESS

—November 1, 1961—

You have asked for our platform:

We are not an organization, therefore we have no platform as such.

We are united around one theme: "End the Arms Race—Not the Human Race."

President Kennedy has said: "Mankind must put an end to war or war will put an end to mankind." We prefer the first alternative.

We're not politicians—we're housewives and working women. We don't make foreign policy—but we know to what end we want it made: toward the preservation of life on earth. The immediate threat to *all* life on earth (yours and mine, the Russians and the Germans, the Indians and the Burmese) is nuclear warfare.

Therefore, we urge all governments, in particular the United States and the Soviet Union, to:

1. Agree *at once* to a ban on *all* atomic weapons testing.
2. Begin *at once* to negotiate in good faith to put all atomic weapons under control of an international agency.
3. Begin *at once* to take concrete steps toward world-wide disarmament.
4. Devote as much of the national budgets, *at once*, to preparation for peace as is now being spent in preparation for war.
5. Declare a moratorium *at once* on name-calling on both sides. Use the United Nations, the press and all mass media for facts, not propaganda.
6. Develop the ability of the United Nations to keep the peace and promote world law.

We intend to work for these goals in many ways. After all, if we don't solve this one fast, all the other good causes will be irrelevant—school aid, renewal of cities, equal opportunity—all will be dust, along with our hopes and fears and dreams—and us.

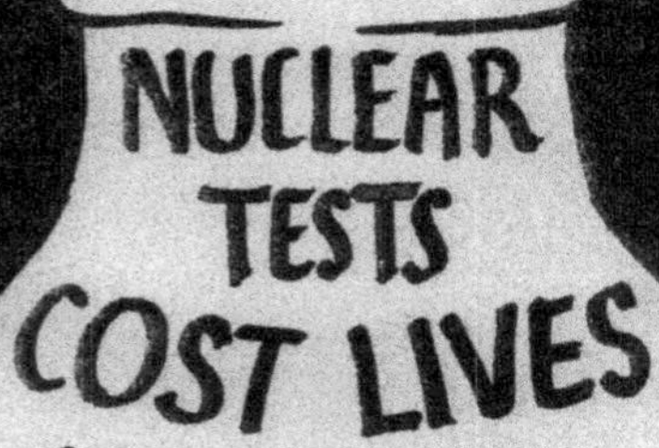

Our children are paying
the price of atomic testing –
in cancer, leukemia, deformities.

RADIOACTIVE IODINE 131

— appears right after tests
— can cause thyroid cancer
— is 18 times more dangerous to babies than to adults

• • • • •

PROTEST –
To show your concern about testing

PROTECT –
to lessen the risk of thyroid cancer

DO NOT USE FRESH MILK

FOR 8 DAYS AFTER EVERY TEST

STOCK UP NOW ON CANNED AND POWDERED MILK
TO MEET YOUR FAMILY'S NEEDS

Women Strike for Peace
1822 Mass. Ave NW, Wash. 6, D.C.

• •

Dear Neighbor:

This message is being sent to you by women of your own community.

American mothers have become deeply concerned about the radioactive contamination of foods. This spring, new highs of radiation are expected from the past Russian tests, according to the AEC and other authorities. And if the U.S. resumes testing, there will be even more danger.

Machinery has recently been developed by the government which can remove most of the radioactive strontium from milk. At least two milk companies have installed such equipment. This clean milk would afford our children a safe supply of calcium, which would help offset the strontium uptake from other foods which cannot be decontaminated.

So little is known that nobody can predict the exact hazard of fallout radiation—but we do know that radioactive strontium *can* cause bone cancer and leukemia. Even though a relatively small number of children may be affected, it is a risk which need not be taken when a preventive measure is available. So we are urging dairies to take the initiative and act now, before the danger grows.

Therefore, we ask if you will—

> Sign the first coupon and put it in your milk bottle or send it to your dairy.
>
> Sign the second coupon and return it to your local contact, so it can be sent to Secretary of [Health, Education, and Welfare Abraham] Ribicoff.
>
> Write the President, urging him to refrain from atmospheric testing. (He reads every 50th letter, and the other [*sic*] are tabulated.)

Yours for Tomorrow!

Women Strike for Peace

To My Milk Company:

As you may know, the nation's milk supply will become increasingly contaminated this year. This can affect the health of our children. We urge you to take action to assure us of a radiation-clean milk supply as soon as possible. Otherwise, I must seriously consider curtailing the use of fresh milk by my family.

Sincerely,

To Secretary of Health, Education, and Welfare Ribicoff:

I am seriously concerned about the hazard to the health of my children from radioactive contamination of foods. In view of the heavy fallout expected this year from past tests, and from tests to come, I strongly urge you to take action now to insure the installation of decontamination equipment by milk companies as a first step in protection. I further urge the creation of a Radiation Protection Agency with the authority to undertake more detailed research and protective action covering all sources of radiation.

Sincerely,

ATOMIC ENERGY COMMISSION

ANNUAL REPORT, 1962

By the early 1960s the United States seemed no closer to the goal of viable commercial nuclear power than it had been at the end of World War II. The appropriate balance between government and industry needed to foster this development remained unclear; should the government take the lead or let private corporations develop commercial applications of nuclear energy? Who should pay initial research and development costs? In 1962 the AEC included a special section in its annual report addressing this issue and discussing the prospects for commercial nuclear power. What role do the authors of the report see for the federal government, and for the AEC in particular? What do they see as the benefits of federal involvement, and of developing commercial nuclear power? Critics of the AEC charged that a single agency should not simultaneously promote and regulate nuclear power. What evidence of this conflict can be found in the report?

THE NEED FOR NUCLEAR POWER

Our technological society requires ample sources of energy. Although large, the supplies of fossil fuels are not unlimited and, furthermore, these materials are especially valuable for many specific purposes such as transportation, small isolated heat and power installations, and as

Excerpted from U.S. Atomic Energy Commission, *Annual Report to Congress of the Atomic Energy Commission for 1962* (Washington, D.C.: U.S. Government Printing Office, 1963), 109–17.

sources of industrial chemicals. Reasonable amounts should be preserved for future generations.

Comparison of estimates of fossil fuel resources with projections of the rapidly increasing rate of energy consumption predict that, if no additional forms of energy were utilized, we would exhaust our readily available lowcost fossil fuels in a century or less and our presently visualized total supplies in about another century. In actual fact, long before they become exhausted we will be obliged to taper off their rate of use by supplementing them increasingly from other sources. . . .

The use of nuclear energy for electric power and, less immediately, for industrial process heat and other purposes is technically feasible and economically reasonable. In addition to its ultimate importance as a means of exploiting a large new energy resource, nuclear electric power holds important near-term possibilities: as a means of significantly reducing power generation costs, especially in areas where fossil fuel costs are high; as an important contributor to new industrial technology and to our technological world leadership; as a significant positive element in our foreign trade; and, potentially, as a means of strengthening our national defense.

In view of the above we have concluded that: *Nuclear energy can and should make an important and, ultimately, a vital contribution toward meeting our long-term energy requirements, and, in particular, that: The development and exploitation of nuclear electric power is clearly in the near- and long-term national interest and should be vigorously pursued.*

THE ROLE OF THE FEDERAL GOVERNMENT

The technological development of nuclear power is expensive. The reactors are complex, and operating units, even of a scaled-down test variety, must of necessity be large and costly. Furthermore, nuclear power does not meet a hitherto unfilled need but must depend for marketability on purely economic advantages that will return the development investment slowly. Hence, the equipment industry could not have afforded to undertake the program by itself. The Government must clearly play a role.

An early objective should be to reach the point where, with appropriate encouragement and support, industry can provide nuclear power installations of economic attractiveness sufficient to induce utilities to install them at their own expense. Once this is achieved the Government should devote itself to advanced developments designed to meet long-range objectives, leaving to industry responsibility for nearer-term improvements. Gradually, as technological maturity is reached, the transition to industry should become complete.

Thus, *the proper role of Government is to take the lead in developing and demonstrating the technology in such ways that economic factors will promote industrial applications in the public interest and lead to a self-sustaining and growing nuclear power industry.* . . .

In our opinion, economic nuclear power is so near at hand that only a modest additional incentive is required to initiate its appreciable early use by the utilities. Should this occur the normal economic processes would, we feel, result in expansion at a rapid rate. The Government's investment would be augmented manyfold by industry. Equipment manufacturers could finance major technical developments, thus reducing the future need for Government participation. . . .

More generally, the introduction of nuclear power technology on a significant scale would add to the health and vigor of our industry and general economy. Technical progress would assist the space and military programs and have other ancillary benefits. Our international leadership in the field would be maintained, with benefit to our prestige and our foreign trade. Nuclear power could also improve our defense posture; it would not burden the transportation system during national emergencies; furthermore, the "containment" required for safety reasons would, if desired, be achieved at little, if any, extra cost by underground installations, thus "hardening" the plants against nuclear attack.

A substantially lesser program would sharply reduce these benefits. Too great a slowdown could result in losing significant portions of industry's present nuclear capability thereby seriously delaying the time at which it would assume a major share of the development costs.

JOHN F. KENNEDY

"COMMENCEMENT ADDRESS AT AMERICAN UNIVERSITY," 1963

The election of John F. Kennedy to the presidency in 1960 brought Cold War tensions to a peak. Kennedy made the decline of American prestige and military preparedness a cornerstone of his campaign, lamenting a purported "missile gap" with the Soviet Union. Once in office, he initiated an unprecedented peacetime buildup of the U.S. military. These moves brought two tense confrontations with the Soviet Union: the Berlin Crisis of 1961 and the Cuban Missile Crisis of 1962. Most experts consider the latter to be the closest that the two superpowers came to a nuclear confrontation, as Kennedy and his advisers gave serious thought to the use of nuclear weapons. The confrontation became a turning point in Cold War and nuclear history. Six months after the Cuban Missile Crisis, Kennedy gave the following address at the graduation ceremony at American University. Why does he call for peace? How does he position the United States relative to the Soviet Union? How does he hope to transcend Cold War rhetoric? What might bring the two nations together to overlook their differences?

. . . .

What kind of peace do we seek? Not a Pax Americana enforced on the world by American weapons of war. Not the peace of the grave or

---

Excerpted from John F. Kennedy, "Commencement Address at American University," June 10, 1963, *American Presidency Project*, www.presidency.ucsb.edu/ws/?pid=9266.

the security of the slave. I am talking about genuine peace, the kind of peace that makes life on earth worth living, the kind that enables men and nations to grow and to hope and to build a better life for their children—not merely peace for Americans but peace for all men and women—not merely peace in our time but peace for all time.

I speak of peace because of the new face of war. Total war makes no sense in an age when great powers can maintain large and relatively invulnerable nuclear forces and refuse to surrender without resort to those forces. It makes no sense in an age when a single nuclear weapon contains almost ten times the explosive force delivered by all of the allied air forces in the Second World War. It makes no sense in an age when the deadly poisons produced by a nuclear exchange would be carried by wind and water and soil and seed to the far corners of the globe and to generations yet unborn. . . .

Let us focus instead on a more practical, more attainable peace—based not on a sudden revolution in human nature but on a gradual evolution in human institutions—on a series of concrete actions and effective agreements which are in the interest of all concerned. There is no single, simple key to this peace—no grand or magic formula to be adopted by one or two powers. Genuine peace must be the product of many nations, the sum of many acts. It must be dynamic, not static, changing to meet the challenge of each new generation. For peace is a process—a way of solving problems. . . .

No government or social system is so evil that its people must be considered as lacking in virtue. As Americans, we find communism profoundly repugnant as a negation of personal freedom and dignity. But we can still hail the Russian people for their many achievements—in science and space, in economic and industrial growth, in culture and in acts of courage. . . .

Today, should total war ever break out again—no matter how—our two countries would become the primary targets. It is an ironic but accurate fact that the two strongest powers are the two in the most danger of devastation. All we have built, all we have worked for, would be destroyed in the first 24 hours. And even in the cold war, which brings burdens and dangers to so many countries, including this Nation's closest allies—our two countries bear the heaviest burdens. For we are both devoting massive sums of money to weapons that could be better devoted to combating ignorance, poverty, and disease. We are both caught up in a vicious and dangerous cycle in which suspicion on one side breeds suspicion on the other, and new weapons beget counterweapons.

In short, both the United States and its allies, and the Soviet Union and its allies, have a mutually deep interest in a just and genuine peace and in halting the arms race. Agreements to this end are in the interests of the Soviet Union as well as ours—and even the most hostile nations can be relied upon to accept and keep those treaty obligations, and only those treaty obligations, which are in their own interest.

So, let us not be blind to our differences—but let us also direct attention to our common interests and to the means by which those differences can be resolved. And if we cannot end now our differences, at least we can help make the world safe for diversity. For, in the final analysis, our most basic common link is that we all inhabit this small planet. We all breathe the same air. We all cherish our children's future. And we are all mortal. . . .

All this is not unrelated to world peace. "When a man's ways please the Lord," the Scriptures tell us, "he maketh even his enemies to be at peace with him." And is not peace, in the last analysis, basically a matter of human rights—the right to live out our lives without fear of devastation—the right to breathe air as nature provided it—the right of future generations to a healthy existence?

While we proceed to safeguard our national interests, let us also safeguard human interests. And the elimination of war and arms is clearly in the interest of both. No treaty, however much it may be to the advantage of all, however tightly it may be worded, can provide absolute security against the risks of deception and evasion. But it can—if it is sufficiently effective in its enforcement and if it is sufficiently in the interests of its signers—offer far more security and far fewer risks than an unabated, uncontrolled, unpredictable arms race.

The United States, as the world knows, will never start a war. We do not want a war. We do not now expect a war. This generation of Americans has already had enough—more than enough—of war and hate and oppression. We shall be prepared if others wish it. We shall be alert to try to stop it. But we shall also do our part to build a world of peace where the weak are safe and the strong are just. We are not helpless before that task or hopeless of its success. Confident and unafraid, we labor on—not toward a strategy of annihilation but toward a strategy of peace.

DAVID E. LILIENTHAL

CHANGE, HOPE, AND THE BOMB, 1963

David Lilienthal served as the first chair of the Atomic Energy Commission (AEC), holding the position from 1946 to 1950. He oversaw the transition of the U.S. nuclear program from a top-secret military operation to a civilian-controlled public endeavor charged with developing both military and commercial nuclear energy. Through his long career as a government administrator, and later as a successful businessman and consultant, Lilienthal remained a staunch proponent of nuclear energy. At times, however, he directed sharp criticism at the way that the federal government and the AEC managed the program. *Change, Hope, and the Bomb* appeared shortly after the Cuban Missile Crisis. How might that crisis have shaped Lilienthal's thinking? Does he view nuclear technology as a sign of progress, or a threat to it? Does he challenge the nuclear consensus, or support it? What has to happen for nuclear technology to lead Americans, and the human species, toward a hopeful future?

. . . .

In the whole of history, no single force has cast a greater terror over all mankind than the Atom. Since Hiroshima, the image of final catastrophe has seized on the minds and hearts of men. The Atom has so heightened the desperate problems of a world in turmoil that it seems

---

Excerpted from David E. Lilienthal, *Change, Hope and the Bomb* (Princeton, N.J.: Princeton University Press, 1963). Pp. 9–10, 18–20, 148–49, 160–62, 167–68.

to have fused them all together into one single universal issue: can an atomic holocaust somehow be avoided?

We have become possessed by our fear. In our efforts to find ways of coping with the threat of the Atom, we have concentrated almost exclusively on the Atom itself. We may from time to time remind ourselves that the problem of the Atom is essentially an intensification of the age-old human problem of managing to live peacefully on earth, but in fact our approach to the Atom has been largely influenced by our fascinated terror of the Bomb. We have tended to regard it as if it were something outside humanity, a satanic power that dominates the affairs of men without actually being part of them. . . .

But I do not believe that any approach can be new or fruitful if it is rooted in the assumption that the Atom is a thing apart from the life of mankind. My conviction is that there is no Atomic Issue; rather, there are Human Issues, and therefore we must look not narrowly at the Atom alone but at the Atom as it is a part of the world today. . . .

The place of the Atom in the life of the world cannot be understood, much less dealt with creatively in its military as well as its nonmilitary aspects so long as we continue to think of it as the exclusive domain of the experts, fragmented and compartmentalized into a score or more fields of expertise. The basis for understanding atomic energy in the life of mankind is not the mastery of physics or abstract mathematics or diplomacy or military science or the exotic nuances of some disarmament plan. To understand the Atom we must reassert what through the ages men have come painfully to know of the condition of man in a changing world, how human affairs are conducted—and misconducted.

The Atom has had us bewitched. It was so gigantic, so terrible, so beyond the power of imagination to embrace, that it seemed to be the ultimate fact. It would either destroy us all or it would bring about the millennium. It was the final secret of Nature, greater by far than man himself, and it was, it seemed, invulnerable to the ordinary processes of life, the processes of growth, decay, change. Our obsession with the Atom led us to assign to it a separate and unique status in the world. So greatly did it seem to transcend the ordinary affairs of men that we shut it out of those affairs altogether; or rather, tried to create a separate world, a world of the Atom.

Doing this, it seems to me, we forgot what we all really know: that if we are to be destroyed, or if some version of the millennium is to be created, man and man alone will be the destroyer or creator, as he has always been. It is still man that counts. Man, always changing, renewing himself from generation to generation, is never obsolete. . . .

We have been following a myth. We have been committed to an illusion about the Atom which has forced us year by year to attempt to make the facts of life fit a concept, instead of trying to frame policies and programs that will fit the facts.

What is the essence of this myth? To my mind, it is this: That because the development of the Atomic Bomb seemed to be the ultimate breakthrough in scientific achievement, in the control of physical matter, we could make a similarly radical departure in dealing with those problems in human affairs which the Bomb so greatly intensified. The Bomb was so colossal, a new force in the world, that we believed a new way must be found to meet its threat, an approach similarly sweeping, similarly radical and worldwide.

In short, our obsession with the Atom drove us to seek a Grand Solution. We became committed to the concept of a total final settlement because nothing short of this would answer the tremendous threat. . . .

Our world simply will not sit still to be fitted for all-embracing theories. To me, therefore, the world of those who prophesy doom or the millennium is not a real world. In the changing world that I know, there are always reasons for deep anxiety—and there are always reasons, too, for hope.

A narrow preoccupation with the Bomb is myopic because it fails to see the potential heights of human achievement, and part of my own sense of hope is based on what I have seen men accomplish. I have seen men achieve the *possible.* I have seen them coping with—not solving, but learning to deal with, to manage—the intricate, contradictory, and refractory problems bound up in their lives as members of a community or a state or a nation or a group of nations. Never have I seen them utterly solve human problems. A basic thesis of this book is that human problems—the great questions of war and peace and also the little questions of family harmony and neighborhood relations—are not susceptible of solution in the sense that scientific questions are solved. Man is too complex, too various, too wonderfully changeable for that.

The concept of Solution is the concept of perfection. It does have a role in the world. From it have sprung the greatest and noblest moral insights and religious ideals. It is certainly not my purpose to deprecate the aspirations which raise up the souls of men.

But by itself, the concept of Solution is an imperfect, even a dangerous, guide. Man is glorious not only because he is capable of formulating great goals, but because he is also capable of moving, achievement by achievement, in their direction. In our country, we have never finally "solved" any basic question. We have never quite lived up to the idealism that animates our Constitution. But being a practical and imaginative people, not easily disheartened by setbacks and disappointments, we have managed, we have coped with our problems successfully enough so that we have not been crushed by poverty, or torn irreconcilably apart by civil disorder, or stagnated by social tensions. . . .

America does have responsibility of leadership in the world, and shoulders that responsibility nobly. In a masterful and mature way under President Kennedy, I think, and his predecessors, Presidents Eisenhower, Truman, and Roosevelt, our government has demonstrated our concern for the whole world, with full support of the American people. America has become indeed the trustee of some great basic principles of freedom.

But this world outlook must not be carried to an extreme that is self-defeating. It is not only charity that begins at home; strength also begins at home, and humaneness, and furthering and *demonstrating at home* the workability of these multiple roads toward peace in the world.

The security not alone of the United States but, to a degree not fully realized, mankind's hopes for peace, depend upon how well we demonstrate to other nations that the physical, cultural, and spiritual foundations of this Republic are steadily being strengthened and developed. The future of people everywhere depends upon how well we show that the concept of a civilized people in one great area of the world, the United States of America, has not been diluted or eroded by either an over-draft on our emotional energies for the noble goal of the well-being of other parts of the world, or as a subconscious escape from the problems at our own doorstep. We must not succumb to the human temptation to turn away from the things we can *do,* and can face up to, that are close to us—in our neighborhood, our home community, or in our country. And one of the most common ways of escaping those close-to-home needs and opportunities to act is to look across the seas or into space for our emotional satisfactions, or as a satisfying distraction from the more prosaic needs at home. . . .

There will be many people who will find the slow diverse process of building a peaceful world quite unappealing and even intolerable while the risk of nuclear destruction hangs over their heads, over everyone's head. Some people will say that they simply cannot face life under the anxieties and tensions of the Atom while the changes I foresee have time to evolve and take effect.

To these I simply know no answer that will be both honest and satisfying. I do know that those who not long ago dug shelters and were preparing to live in them, or those few who ran away to some remote safe place to hide, soon found this to be a less tolerable way of existing than putting the potential danger out of mind. I believe that it is not how long one lives but the quality of life that counts, that sustains men. I do know that man is a most adaptive creature. And I do know that we are not the first generation of human beings who have had to live their whole lives face to face with mortal danger, whether of plague or constant warfare or the forces of nature; and I know too that it has often been those very periods, such as this one, when acute anxiety and danger of death were man's constant companions, that brought forth some of the most creative and robust chapters of history.

I believe in man. I believe he will not perish. Nor will the works of his spirit and brain and imagination vanish from the earth. I believe civilization will ride through this storm.

I do not believe that God created man and endowed him with the capacity to unlock the energy within the very heart of matter in order that he should use that knowledge to destroy this beautiful world, which is the handiwork not of man, but of God.

JOHN F. KENNEDY

"ADDRESS TO THE AMERICAN PEOPLE ON THE NUCLEAR TEST BAN TREATY," 1963

The Cuban Missile Crisis prompted new thinking about the arms race on both sides of the Cold War and fresh climate for negotiation. Although domestic opponents of nuclear testing had begun to question the environmental impacts of radioactivity and fallout, most experts point to other motivations for the agreement, such as the cost of testing, the need to moderate Cold War tensions, and the concern about nuclear proliferation. Over a hundred countries eventually signed the treaty, which banned nuclear testing in the atmosphere, in outer space, and underwater, although not underground. How does President Kennedy describe the agreement? What, according to Kennedy, motivated this agreement? What role do environmental concerns play? How does he propose to get outside the cycle of Cold War tension and response?

I speak to you tonight in a spirit of hope. Eighteen years ago the advent of nuclear weapons changed the course of the world as well as the war. Since that time, all mankind has been struggling to escape from the darkening prospect of mass destruction on earth. In an age when both sides have come to possess enough nuclear power to destroy the human race several times over, the world of communism and the world of free

---

Excerpted from John F. Kennedy, "Radio and Television Address to the American People on the Nuclear Test Ban Treaty," July 16, 1963, Miller Center, University of Virginia, http://millercenter.org/president/speeches/detail/3377.

choice have been caught up in a vicious circle of conflicting ideology and interest. Each increase of tension has produced an increase of arms; each increase of arms has produced an increase of tension.

In these years, the United States and the Soviet Union have frequently communicated suspicion and warnings to each other, but very rarely hope. Our representatives have met at the summit and at the brink; they have met in Washington and in Moscow; in Geneva and at the United Nations. But too often these meetings have produced only darkness, discord, or disillusion.

Yesterday a shaft of light cut into the darkness. Negotiations were concluded in Moscow on a treaty to ban all nuclear tests in the atmosphere, in outer space, and under water. . . .

But the achievement of this goal is not a victory for one side—it is a victory for mankind. It reflects no concessions either to or by the Soviet Union. It reflects simply our common recognition of the dangers in further testing.

This treaty is not the millennium. It will not resolve all conflicts, or cause the Communists to forego their ambitions, or eliminate the dangers of war. It will not reduce our need for arms or allies or programs of assistance to others. But it is an important first step—a step towards peace—a step towards reason—a step away from war. . . .

A war today or tomorrow, if it led to nuclear war, would not be like any war in history. A full-scale nuclear exchange, lasting less than 60 minutes, with the weapons now in existence, could wipe out more than 300 million Americans, Europeans, and Russians, as well as untold numbers elsewhere. And the survivors, as Chairman [Nikita] Khrushchev warned the Communist Chinese, "the survivors would envy the dead." For they would inherit a world so devastated by explosions and poison and fire that today we cannot even conceive of its horrors. So let us try to turn the world away from war. Let us make the most of this opportunity, and every opportunity, to reduce tension, to slow down the perilous nuclear arms race, and to check the world's slide toward final annihilation. . . .

[This] treaty can be a step towards freeing the world from the fears and dangers of radioactive fallout. Our own atmospheric tests last year were conducted under conditions which restricted such fallout to an absolute minimum. But over the years the number and the yield of weapons tested have rapidly increased and so have the radioactive hazards from such testing. Continued unrestricted testing by the nuclear powers, joined in time by other nations which may be less adept in limiting pollution, will increasingly contaminate the air that all of us must breathe.

Even then, the number of children and grandchildren with cancer in their bones, with leukemia in their blood, or with poison in their lungs might seem statistically small to some, in comparison with natural health hazards. But this is not a natural health hazard—and it is not a statistical issue. The loss of even one human life, or the malformation of even one baby—who may be born long after we are gone—should be of concern to us all. Our children and grandchildren are not merely statistics toward which we can be indifferent.

Nor does this affect the nuclear powers alone. These tests befoul the air of all men and all nations, the committed and the uncommitted alike, without their knowledge and without their consent. That is why the continuation of atmospheric testing causes so many countries to regard all nuclear powers as equally evil; and we can hope that its prevention will enable those countries to see the world more clearly, while enabling all the world to breathe more easily. . . .

But now, for the first time in many years, the path of peace may be open. No one can be certain what the future will bring. No one can say whether the time has come for an easing of the struggle. But history and our own conscience will judge us harsher if we do not now make every effort to test our hopes by action, and this is the place to begin. According to the ancient Chinese proverb, "A journey of a thousand miles must begin with a single step."

My fellow Americans, let us take that first step. Let us, if we can, step back from the shadows of war and seek out the way of peace. And if that journey is a thousand miles, or even more, let history record that we, in this land, at this time, took the first step.

Thank you and good night.

PART 4

CONFRONTING PARADOX

The Partial Nuclear Test Ban Treaty of 1963 helped lessen anxieties about the prospect of nuclear war, and the 1960s brought an ebb in nuclear fear. American and Soviet foreign policies moved toward the stance of détente—a deliberate lessening of tensions between the superpowers. The peace movement concentrated less on nuclear weapons and increasingly on the Vietnam War. Concerns about nuclear energy focused instead on its nonmilitary uses, in particular on commercial nuclear power.

After years of false starts, commercial nuclear power finally established a place in the nation's energy portfolio in the late 1960s. The expense of constructing and operating nuclear generators had scared away many utility companies, but as concerns about the environmental costs and limited supply of fossil fuels grew, nuclear technologies seemed more attractive. As the utilities invested in reactors, public concern about the risks associated with nuclear power provoked an antinuclear movement focused more on power production than on weapons. Mounting public opposition and distrust, and also the 1973 oil crisis, prompted Congress to split the Atomic Energy Commission into two agencies: the Nuclear Regulatory Commission and the Energy Research and Development Administration, which had authority over research, development, and promotion of civilian nuclear power. Public opposition, slowing economic growth (and demand for electricity), and an increasingly contentious regulatory environment impeded the further development of commercial nuclear power. By the late 1970s the industry faced an uncertain future.

The documents in part 4 explore the shift in the focus of the conversation about nuclear energy from weapons to electricity generation.

Americans wrestled with a different aspect of the nuclear paradox, replacing discussions of life and death in nuclear warfare with conversations about how to balance risk and reward in the pursuit of progress and affluence. Environmental concerns moved to the center of the debate. What kinds of assumptions did stakeholders make about economic growth, environmental health, and definitions of progress? How did they seek to balance the economic and environmental costs and benefits of nuclear energy?

GLENN T. SEABORG

ENVIRONMENTAL EFFECTS OF PRODUCING ELECTRIC POWER, 1969

During his illustrious career chemist Glenn Seaborg touched almost every aspect of American encounters with nuclear energy. He worked with the Manhattan Project during World War II, earned a Nobel Prize for the discovery of plutonium and other elements, served as the chair of the Atomic Energy Commission from 1961 to 1971, and stood as a prominent advocate for both arms control and the peaceful applications of nuclear technology. In the congressional testimony below, Seaborg describes nuclear power as an appropriate reaction to the emerging environmental movement. How does he come to his conclusions? Does he accept the claims of environmentalism or challenge them? In what ways does he see nuclear power as a response to these concerns? How does Seaborg position himself relative to the nuclear consensus?

....

If public interest is the criterion, then timely is exactly the word to use in describing these hearings on the Environmental Effects of Producing Electric Power. One would have to be totally cut off from civilization these days—or both blind and deaf—not to be fully aware of the public's concern with what has been broadly termed The Environment. There is hardly a day in the week, or an hour in the day, when one does not see a newspaper or magazine article, hear a radio program or

---

Excerpted from Glenn T. Seaborg, Joint Committee on Atomic Energy, *Environmental Effects of Producing Electric Power, Part 1*, 91st Congress, 1st session, 1969, 84–89.

view a TV show in which the subject of pollution—in terms of waste, water, air, chemical, noise or other varieties—is brought up in some way. It was pointed out to me the other day that in a Sunday edition of a nationally known newspaper, three sections alone had eight articles with 2,100 lines devoted to environmental effects. And this is typical of what we might call "the pollution press coverage" we are receiving from all media.

I submit that on an overall basis this public awareness and alarm is a very good thing. Man in general has always had a tendency towards excesses and indiscretions. Invariably such faults, whether they are manifested in matters involving his own personal life and health, or in those expressed through his society (which ultimately also affect the individual's life and health), have a way of surfacing sooner or later. . . .

Now, while I have emphasized the public's awareness and alarm over environmental conditions, I must express my own alarm, or at least my extreme concern, over a related matter.

I am concerned that, for all the extensive coverage of pollution, much of the public is being ill-informed and misinformed about many environmental matters. I am concerned that for every bit of valid criticism there is more than an equal amount of unsubstantiated fear mongering. Again we are faced with a matter of excess, or at least with an imbalance, where alarm and a sense of urgency is present in abundance, but where the information, funds, time, and spirit of cooperation—all so necessary to constructive action—are not readily at hand. . . .

I think this situation must change if we are to deal successfully with our environmental problems. We must have less hysteria, less searching for scapegoats, less polarization of conservationists and technologists, and less conflict between those engaged in the various disciplines that affect and deal with the environment. We need a more tempered sense of urgency, more knowledge, more cooperation, and much more of the positive outlook and approach.

The degrading of our environment has not been the fault of any one group or element of society. Nor will its future be determined by the action of any one segment, industry, Government, or the public. It is a task for all—and it should be. We all breathe the same air that forms such a thin and precious envelope around this unique planet. We all share the same meager 2 percent of fresh water on its surface. We all need to use the same limited resources and space this earth provides. And we all want to turn this same earth over to our children, and their children, as a clean, livable, and attractive home. I strongly believe that we can do this—not by complaining about what has been, or even what is, but by exercising a little wisdom and a lot of hard work toward what can be. . . .

To begin with, most of us recognize today that our total environment is a close combination of both our natural world and the technological civilization we have built. Contrary to what some people are saying these days, I believe that both elements of this combination are necessary and desirable and, what's more important here, can be highly compatible. Essential to this is the constructive use of energy—energy that is readily available, abundant, economic, and that can be applied massively but with a minimum of impact on the environment. We must also face the fact that a growing world population with both rising standards of living and volatile rising expectations will demand a huge amount of power in the years ahead. . . .

There is no doubt that a large amount of this energy, particularly in terms of oil for the transportation and power fields, is going to come from fossil fuels over the next few decades. But in spite of forecasts of large reserves, we know that these natural resources are limited. We also know that there is a limit to nature's ability, and our own human tolerance, to absorb all the pollution that would result should we try to burn up all these fuels over the next century or so. I am not going to document the pollution loads that would accrue from the combustion of that massive amount of fuel. I think the members of this committee have many times been made aware of these amounts and their consequences. Fortunately, we have for the generation of electric power an alternate fuel—nuclear energy.

As I have said on other occasions, I believe that nuclear energy has arrived on the scene—historically speaking—in the nick of time. I base this belief on several factors:

1. The projected demand for power based on population growth and increasing per capita consumption of electricity. . . .
2. The need for a cleaner and more manageable source of energy to reduce the degradation of the environment.

3. The need also for abundant and very economic energy in a world of diminishing natural resources where such energy may well determine how many people can be supported and at what living standard....

Now, since we are eventually going to live in a world that will have to depend on the energy of the atom, we must learn to live with the atom wisely. This means we must recognize, anticipate and deal with all the environmental aspects and prospects of nuclear energy. I believe we are doing this, and doing it well. This type of technological development is something that has never before been attempted in the history of man. No technology has been born and developed with the regard for human safety and well-being that is inherent in the development of nuclear energy.

In fact, you might say that the extent of our knowledge about the biological aspects of nuclear energy has been a problem to us—or at least to those in the nuclear field who are impatient. The tremendous amount of knowledge we have accumulated over some 20-odd years has made us almost overly conservative in the development of nuclear power. I have often thought that if the potential health and safety implications of so many aspects of our lives—our chemical products, our foods, our transportation systems, our athletic activities, even our sleeping habits, to name a few—were so well-known and documented, we would have a very apprehensive public—literally afraid to eat or drink anything or go anywhere or do anything.

Fortunately, because of our knowledge of nuclear energy, and the way we have developed it in accordance with that knowledge, we have at hand a unique opportunity to advance an abundant source of power with a minimum of environmental impact. We are following such a course, fulfilling such an opportunity....

[. . . All] that I have seen and heard, my total experience in the nuclear energy field for more than a quarter of a century and my association with others who have devoted their lives to this field, has given me the firm conviction that the environmental problems associated with nuclear energy are manageable. With good planning and continued dedicated work on the part of those in the nuclear field, our electric utilities and those Government agencies that regulate our Nation's power systems, we can have safe, clean and reliable nuclear power—as much of it as we will need.

The suggestions I have made earlier of vast benefits that can be derived for mankind from nuclear energy were not made without full awareness that there are inherent in this technology certain risks and potential hazards to health and safety, as there are risks in many other activities. Recognition of the fact of these risks is the basic reason for the comprehensive system of safety review and regulatory controls set up by the Congress for the protection of people and the safety of the reactor facilities and for the extensive programs of safety research in both the physical and biological aspects of nuclear power plants and radiation.

In spite of the current wave of misunderstanding and fear registered by a certain segment of the public, I think we are going to prove this important point—that the benefits related to nuclear power will outweigh the risks involved by a factor far greater than most of our modern technologies can boast.

There will be continuous agitation, there will be adjustment and compromise—more important, there will be more understanding and a better working relationship between reasonable and rational environmentalists and technologists who will see that they are not as far apart as they believe. As a result, we will see in the long run more nuclear power and a healthier environment.

When we have gotten past this point, I believe we are going to see some remarkable things happening with nuclear power. We will find, with good site planning and the esthetic designing of nuclear plants, that nature and technology are not incompatible. . . . We could see the use of abundant and very economic nuclear energy having a widespread beneficial effect on many other environmental problems—helping us to supplement and control our water resources, helping us to recycle much of our solid waste, thus preserving our diminishing mineral resources and eliminating many eyesores and environmental blights on our landscape. And we will ultimately see this kind of nuclear energy having a remarkable effect on world development, helping to lift billions of energy-starved individuals into the mainstream of the 20th century.

Perhaps the most disturbing thing about the current reaction to environmental problems is the attitude it is engendering—a fear that is making many look backwards. There are some people whose only reaction to the possibility of future power shortages—and "blackouts" and "brownouts"—should we fail to plan and build now to meet our future needs, is that we should reduce our use of electricity, turn out our lights. There are others who are so irrational in their fear of nuclear power, and so desperate for alternatives, that they have seriously advocated harnessing the Gulf Stream, or icebergs, or volcanoes, or hot air balloons. Fortunately, most people are not willing to sit in the dark, or search in the dark for a better life for themselves and their children.

We who are involved in developing nuclear power to provide for future electricity needs are naturally disturbed by that public resistance which seeks to halt or slow down such development. However, along with our obligation to safeguard the natural environment we also have a responsibility to help supply our people with the power to run a technologically sustained society. In the years ahead, today's outcries about the environment will be nothing compared to cries of angry citizens who find that power failures due to a lack of sufficient generating capacity to meet peak loads have plunged them into prolonged blackouts—not mere minutes of inconvenience, but hours—perhaps days—when their health and well-being and that of their families, may be seriously endangered.

MINNESOTA ENVIRONMENTAL
CONTROL CITIZENS ASSOCIATION

"A NUCLEAR ENERGY GAMBLE," CA. 1969

By the late 1960s it appeared that developing technologies and favorable market prospects pointed to the long-delayed arrival of the commercial nuclear power industry. As construction of new plants began, an antinuclear movement slowly took shape, usually on a site-by-site basis. Instead of a national movement, local residents in affected communities voiced their concerns about the environmental impact and safety of commercial reactors. One of the first such controversies arose in Minneapolis–Saint Paul, where a large utility company planned to build two nuclear reactors on the Mississippi River—the Monticello plant upstream from the Twin Cities and the Prairie Island plant downstream. Early in the process, the Minnesota Environmental Control Citizens Association (MECCA) mobilized to fight against the location and planned regulation of radioactive discharges at the Monticello plant. How does MECCA try to raise concern and awareness about the issue? How does its characterization of the environmental implications of nuclear power differ from that of Glenn Seaborg?

Excerpted from Minnesota Environmental Control Citizens Association (MECCA), Anti-Nuclear Pamphlet, ca. 1969, Minnesota Environmental Control Citizens Association Papers, Box 3, Folder "Monticello, undated, 1969," Minnesota Historical Society, Saint Paul, Minn. Courtesy of the Minnesota Historical Society.

MONTICELLO

A NUCLEAR ENERGY GAMBLE

The stakes: mutation, cancer, death

Citizens are concerned about the idea of radioactive wastes being dumped into the Mississippi at Monticello. We should be. It's our drinking water. And in spite of the assurances of safety from the Atomic Energy Commission and Northern States Power Company—the safety and performance records of nuclear energy plants have been dismal.

Of the original 12 nuclear power plants that have been put into operation, 8 have failed—including the one at Elk River where radioactive leaks forced shutdown—and the Northern States Power "Pathfinder" plant in Sioux Falls which exceeded its yearly concentration limit despite being operated below full power. Three plants have been abandoned (one at an estimated $7 million decontamination cost, paid by the taxpayer, of course).

In all cases where these plants failed, citizens had been assured, as now, of complete safety.

Q. If there were a real danger to health from radioactive waste, would the Atomic Energy Commission approve of such a plant?

A. It appears that the AEC not only would but in fact has approved of such plants. The Hanford, Washington Atomic Energy facility on the Columbia River is an example.

A 1965 study showed that Oregon counties bordering the Columbia River downstream from the Hanford facility had a 53 percent higher cancer rate than the rest of the state. The JOURNAL OF ENVIRONMENTAL HEALTH reported: "This physiographic pattern of malignancy provides strong circumstantial evidence that not just leukemia but all types of cancer are influenced by bodily ingested radioisotopes in quantities heretofore thought safe." We might add, "declared safe" by the AEC.

Q. But why would the AEC approve a nuclear power installation where even the slightest question of safety exists?

A. It is important to keep in mind that the AEC was established to promote the use of nuclear energy. Limiting such use, even for safety reasons, is clearly a conflict of interest for the AEC.

Q. What is a "safe level" of radioactivity in the environment?
A. There is no "safe level" of radioactivity. Radiation as minimal as X-ray exposure of an unborn child is associated with leukemia in later life. Standards depend on how many deaths and mutations we are willing to accept.

For example, the Federal Radiation Council has set its standards at .5 rem yearly exposure. "If we assume the population of the Twin Cities metropolitan area to be two million, then a continuing yearly exposure of .5 rem—the FRC standard dose—would be expected to cause from 10 to 100 cases of leukemia per year and about an equal number of other types of neoplasms (cancer) . . . Whether a loss of this magnitude is acceptable to society can only be determined by considering the benefits to be gained from a particular use of atomic energy."

A question one might ask is "whose benefits and whose deaths?"

Q. How much radioactive waste would the proposed Monticello Plant discharge into the Mississippi?
A. Northern States Power estimates a total waste, including fuel leaks, of 91.4 Curies yearly.

General Electric, who has a reputation for seriously underestimating radioactive discharge, guesses 30,000 Curies the first year. Note the discrepancy: 29,998.6 Curies. *The real figure is anybody's guess. . . .*

Q. What about the present argument between Northern States Power and the Pollution Control Agency as to allowable limits of radioactive contamination?
A. This is a sham battle diverting attention from the real point that *no* amount of radioactive waste is safe and under *no* conditions should dumping it in our drinking water be tolerated.

Eugene P. [Odum], in his widely used textbook, FUNDAMENTALS OF ECOLOGY, says: "Should a system receive a higher level of radiation than that under which it evolved, nature will not take it 'lying down,' so to speak; adaptations and adjustments will occur along with elimination of sensitive strains or species."

Put another way: radioactive waste dumped into the Mississippi will result in mutations or freaks in plants, animals, fish and people. Cancer and the death rate due to cancer will increase. No limits have been set on the increase of illness and death that is "acceptable." That will apparently depend on how loud people protest as they learn what is happending [*sic*]. . . .

Q. What can you do?
A. Make your voice heard. Don't leave it to the other guy. Protest now against dumping radioactive waste in *any* amount into the Mississippi River or any other body of water in Minnesota.

LENORE MARSHALL

"THE NUCLEAR SWORD OF DAMOCLES," 1971

In the title of this article, poet and peace activist Lenore Marshall—one of the founders of SANE (the National Committee for a Sane Nuclear Policy) in 1957—calls nuclear energy a "Sword of Damocles," referring to an ancient Greek story with the moral that fortune and power also bring danger and responsibility. Marshall published this piece in *The Living Wilderness*, the magazine produced by the Wilderness Society to spread its message about the need to protect wilderness areas from development and the intrusions of modern consumer society. Marshall's brother-in-law, Bob Marshall, served as a founding member and key financial supporter of the society. How does Lenore Marshall relate the threats posed by nuclear energy to other environmental issues, particularly to wilderness protection? To what extent does she see science and technology as part of the problem, or part of the solution, to these concerns? The two scientists that she references—John Gofman and Arthur Tamplin—engaged in a controversial debate with Atomic Energy Commission (AEC) experts over radiation safety and assumed a prominent role in the antinuclear movement for their use of science to question the benefits of nuclear power.

During the years since Hiroshima . . . we have recognized to our sorrow and terror that our entire planet has joined the wilderness in its struggle

---

Excerpted from Lenore Marshall, "The Nuclear Sword of Damocles," *The Living Wilderness* (Spring 1971): 17–19. Courtesy of the Wilderness Society.

for survival; not only the wilderness but the whole world is in peril. Nothing, no matter how remote, is immune. Great tracts of fertile land, plant life and animal life in forests, plains, oceans, rivers, and lakes, have been joined by human life in the danger of extinction. The greatest threat to the continuance of animal, vegetable, and human existence comes from the nuclear sword of Damocles that hangs over our heads.

By great good luck, despite the minor accidents, there has not yet been a massive release. However, since sources of nuclear contamination are proliferating, the chances of a major disaster are also increasing; such a disaster could devastate a number of states and cause thousands more cases of cancer and genetic defects and deaths. There is a fundamental difference between radioactive pollutants and other pollutants such as DDT, NTA [nitrilotriacetic acid], oil, and automobile exhaust. All the latter are stable compounds, and there are possibilities of eliminating them or of rendering them harmless. But radioactive atoms are deranged atoms whose high-energy emissions from the nucleus cannot be stopped or, presto, made innocent by a lawsuit or a wave of a wand; they taper off at their own rate—240,000 years for radioactive plutonium 239, which happens to be a basic element in both the military and peaceful application of nuclear energy.

Cockroaches are said to withstand the effects of radiation quite nicely. Other animals, wild or otherwise, fare worse.

Since there is no way to turn off radioactivity, nuclear pollution is in a class by itself. Therefore, to whatever extent is possible, we must prevent any more of it from occurring.

We are already bearing the legacy of some earlier activities—radium from uranium mine wastes eroding into the Colorado and into other rivers, plutonium 238 in the atmosphere from a misfired navigational satellite (1964), and fallout from the atmospheric nuclear bomb tests. They are all, of course, still with us. For instance some of the radioactive cesium 137 will still be around 300 years from now and radioactive carbon 14 another 57,000 years. The strontium 90 fallout created by atmospheric tests was enough to work its way into the bones of almost every child tested for it in the Northern Hemisphere. . . . Since all radiation exposure is assumed to be harmful, whether it comes from bombs, medical X-rays, nuclear power plants, rocks, or the stars, what counts is the amount we accumulate and which we can still limit. The only hopeful thing to be said about this peril is that it is still possible to control it, keeping doses of radiation to safer permissible levels....

When plutonium 239 falls on the test site in Nevada, the land is fenced off and posted. The problem is how to confine that plutonium to that fenced-off place, against wind and oxidation, for the next 240,000 years—when it will no longer be able to hurt us. Near Denver, Colorado, local scientists have proven that significant amounts of plutonium have escaped from the Rocky Flats plant where warheads are manufactured. After denying the possibility, the A.E.C. has confirmed the findings....

Today's environmental crisis proves that much modern technology now actually functions to the detriment of society. It has become disoriented from society. Science and scientists are not omniscient; in fact many scientists are attached to special interests in government and industry. As Doctors John W. Gofman and Arthur R. Tamplin say of science and technology: "... they create the illusion that if we really get into trouble with our environment, science and technology will be able to rescue us; and they divert the scientific manpower away from more meaningful programs." Thus, within the fact that there are seismic, tidal wave, and radioactive hazards from nuclear weapons-testing underground, there lies the greater danger that weapons-testing is part of a general framework of thinking that war is thinkable.

Doctors Gofman and Tamplin continue: "Science in itself is not bad or good; that is why it has no ethics. Without application, science is meaningless. But most of science in this country is meant to be applied, and hence the government, hand in glove with industry, rules over science by controlling the purse strings.... Quite obviously we need a mechanism for effectively criticizing present day science and technology, and for articulating a new set of priorities that would lead science and technology to fulfilling the needs of society.... They must offer alternative programs that represent routes to the solution of the needs of society."...

It is argued that the country's increased need for electrical power necessitates nuclear plants and that defense needs necessitate further weapons development. As for the latter, since we already have means for overkill beyond that of any other country, and since the continuance of the arms race leads to a deadly tit-for-tat psychology that can only end in catastrophe, the sooner a moratorium on development and accumulation of nuclear weapons is called the safer we shall be. If the world aims at universal disarmament, perhaps elephants and seals and eagles and sparrows and pine trees and fish and roses and children will survive. A moratorium on the burgeoning nuclear reactor business must similarly be called.

What are some of the alternatives? Without nuclear energy would there be brown-outs? Would a million sparklers around advertisements be cut in half? Would the electric carving knife not cut? The answer is that we *can* obtain the power we need. The lights will not go out. Even if this were the case, one must ask which is more important: more lights or life itself? Moreover *there are safer alternatives* to nuclear electricity. There is the further development of fossil fuel, which may be better utilized and made "clean" by means of new processes. . . . Promising work is being done to develop the use of solar energy; it is said that the sun's heat falling on Death Valley alone could solve a multitude of power needs. Certainly, much electric energy that is wasted today could be conserved. . . .

The marvel of our mass society, of our intricate civilization, of our establishments and vast impersonal structures, is that the individual can always do something. The individual has always performed miracles, and he still can. He can save his wilderness, he can save animal and vegetable life, he can save himself. He can understand his predicament, and if he has the will to do so he can take steps to save what he loves; one man—one woman—can start to build a bridge whereon others may walk. Will individuals tackle this new proliferating danger before it is too late?

CALVERT CLIFFS' COORDINATING COMMITTEE, INC., V. UNITED STATES ATOMIC ENERGY COMMISSION, 1971

The emergence of the environmental movement in the late 1960s changed the landscape for commercial nuclear power. In 1969, Congress passed the National Environmental Policy Act (NEPA), which required that all federal agencies consider the environmental impacts of their actions, usually in the form of a comprehensive environmental impact statement. In the *Calvert Cliffs* decision a federal court of appeals deliberated on the extent to which NEPA applied to the Atomic Energy Commission (AEC). The AEC claimed that it had no jurisdiction under NEPA to consider the environmental impacts of the reactors it licensed, arguing that this responsibility fell to other agencies. The court disagreed, however, and its decision handed the antinuclear and environmental lobbies a powerful tool: citizen groups now had the ability to engage more directly in the permitting process and could often slow that process significantly. How do the judges come to their decision? In what ways do they require the AEC to balance environmental and economic factors?

....

These cases are only the beginning of what promises to become a flood of new litigation—litigation seeking judicial assistance in protecting our natural environment. Several recently enacted statutes attest to the commitment of the Government to control, at long last, the destructive

---

Excerpted from *Calvert Cliffs' Coordinating Committee, Inc., v. United States Atomic Energy Commission*, 449 F.2d 1109 (U.S. Court of Appeals, D.C. Court, 1971).

engine of material "progress." But it remains to be seen whether the promise of this legislation will become a reality. Therein lies the judicial role. In these cases, we must for the first time interpret the broadest and perhaps most important of the recent statutes: the National Environmental Policy Act of 1969 (NEPA). We must assess claims that one of the agencies charged with its administration has failed to live up to the congressional mandate. Our duty, in short, is to see that important legislative purposes, heralded in the halls of Congress, are not lost or misdirected in the vast hallways of the federal bureaucracy.

NEPA, like so much other reform legislation of the last 40 years, is cast in terms of a general mandate and broad delegation of authority to new and old administrative agencies. It takes the major step of requiring all federal agencies to consider values of environmental preservation in their spheres of activity, and it prescribes certain procedural measures to ensure that those values are in fact fully respected. Petitioners argue that rules recently adopted by the Atomic Energy Commission to govern consideration of environmental matters fail to satisfy the rigor demanded by NEPA. The Commission, on the other hand, contends that the vagueness of the NEPA mandate and delegation leaves much room for discretion and that the rules challenged by petitioners fall well within the broad scope of the Act. We find the policies embodied in NEPA to be a good deal clearer and more demanding than does the Commission. We conclude that the Commission's procedural rules do not comply with the congressional policy. . . .

NEPA, first of all, makes environmental protection a part of the mandate of every federal agency and department. The Atomic Energy Commission, for example, had continually asserted, prior to NEPA, that it had no statutory authority to concern itself with the adverse environmental effects of its actions. Now, however, its hands are no longer tied. It is not only permitted, but compelled, to take environmental values into account. Perhaps the greatest importance of NEPA is to require the Atomic Energy Commission and other agencies to *consider* environmental issues just as they consider other matters within their mandates. . . .

. . . In general, all agencies must use a "systematic, interdisciplinary approach" to environmental planning and evaluation "in decisionmaking which may have an impact on man's environment." In order to include all possible environmental factors in the decisional equation, agencies must "identify and develop methods and procedures * * * which will insure that presently unquantified environmental amenities and values may be given appropriate consideration in decisionmaking along with economic and technical considerations." "Environmental amenities" will often be in conflict with "economic and technical considerations." To "consider" the former "along with" the latter must involve a balancing process. In some instances environmental costs may outweigh economic and technical benefits and in other instances they may not. But NEPA mandates a rather finely tuned and "systematic" balancing analysis in each instance. . . .

The special importance of the pre-operating license stage is not difficult to fathom. In cases where environmental costs were not considered in granting a construction permit, it is very likely that the planned facility will include some features which do significant damage to the environment and which could not have survived a rigorous balancing of costs and benefits. At the later operating license proceedings, this environmental damage will have to be fully considered. But by that time the situation will have changed radically. Once a facility has been completely constructed, the economic cost of any alteration may be very great. In the language of NEPA, there is likely to be an "irreversible and irretrievable commitment of resources," which will inevitably restrict the Commission's options. Either the licensee will have to undergo a major expense in making alterations in a completed facility or the environmental harm will have to be tolerated. It is all too probable that the latter result would come to pass.

By refusing to consider requirement of alterations until construction is completed, the Commission may effectively foreclose the environmental protection desired by Congress. It may also foreclose rigorous consideration of environmental factors at the eventual operating license proceedings. If "irreversible and irretrievable commitment[s] of resources" have already been made, the license hearing (and any public intervention therein) may become a hollow exercise. This hardly amounts to consideration of environmental values "to the fullest extent possible."

A full NEPA consideration of alterations in the original plans of a facility, then, is both important and appropriate well before the operating license proceedings. It is not duplicative if environmental issues were not considered in granting the construction permit. And it need not be duplicated, absent new information or new developments, at the operating license stage. In order that the pre-operating license review be as effective as possible, the Commission should consider very seriously the requirement of a temporary halt in construction pending its review and the "backfitting" of technological innovations. For no action which might minimize environmental damage may be dismissed out of hand. Of course, final operation of the facility may be delayed thereby. But some delay is inherent whenever the NEPA consideration is conducted—whether before or at the license proceedings. It is far more consistent with the purposes of the Act to delay operation at a stage where real environmental protection may come about than at a stage where corrective action may be so costly as to be impossible.

Thus we conclude that the Commission must go farther than it has in its present rules. It must consider action, as well as file reports and papers, at the pre-operating license stage. . . .

We hold that . . . the Commission must revise its rules governing consideration of environmental issues. We do not impose a harsh burden on the Commission. For we require only an exercise of substantive discretion which will protect the environment "to the fullest extent possible." No less is required if the grand congressional purposes underlying NEPA are to become a reality.

WILLIAM R. GOULD

"THE STATE OF THE ATOMIC INDUSTRY," 1974

In October 1973 the oil-producing nations of the Middle East responded to U.S. support for Israel by enacting an oil embargo. By the spring of 1974 the price of oil had nearly quadrupled, prompting responses ranging from the rationing of gasoline to a rethinking of American energy supply, security, and policy. Some experts promoted conservation and the development of "soft" energy paths like solar power. Others, like engineer and utility executive William R. Gould, thought the oil crisis would boost the market for commercial nuclear power. Gould chaired the industry group Atomic Industrial Forum, and he delivered the following speech at the organization's annual conference. How does he describe the state of the nuclear power industry? What has checked the growth of nuclear power, and how does he want to address these concerns? What does he see as the proper role of the federal government in planning and regulating nuclear power? How does he characterize the different stakeholders in the energy debates?

I've been asked to discuss the State of the Industry this morning, and in short I would suggest that more than ever before it is in a state of uncertainty. All energy industries are sharing this state, to some extent,

---

Excerpted from William R. Gould, "The State of the Atomic Industry: The First Twenty Years, and the Next," William R. Gould Papers, Speech File, Box 42, Folder 6, J. Willard Marriott Library, University of Utah, Salt Lake City. Courtesy of the Special Collections Department, J. Willard Marriott Library, University of Utah.

because of the international politics of petroleum, the state of our economy, the steadily changing energy authority here in Washington, the unpredictable economics of energy in the next few years, and even the uncertainty about future demand curves and the effects of conservation and increased prices. The electric utilities, I know all too well, have the additional problems of rapidly increasing capital and fuel costs, high interest charges, a depressed stock and bond market, and in many cases inadequate rates. All these pressures seem to focus especially on nuclear power, which is the highest capital-cost generating form and the most vulnerable to obstruction, delays and escalation, as a result of its unique licensing process.

Even with all these uncertainties, though—which are hardly news to any of you—it seems clear that the outlook for the nuclear power industry must be the most bullish of any industry that comes to mind. If the nation as a whole pulls out of its slump, nuclear power must be a part of the recovery. . . .

I hardly need to elaborate on the nation's precarious energy supply situation. In the past year, thanks to the world's principal oil-exporting countries, the United States has begun to realize how dependent it is on international politics and whims of other countries for its basic energy resource, petroleum. The unprecedented problems that we saw last winter are barely shadows of the potential crises of the future if we continue to expand our reliance on overseas oil supplies. As many in the industry have been predicting for years, we are in danger of losing not only a significant portion of the energy that fuels our economy, but also our national freedom of action in dealing with international problems. Every expert agrees that domestic petroleum and coal alone cannot keep us out of this trap; the only other alternative available is nuclear power. . . .

This past year has dramatized the seemingly contradictory situation of nuclear power. Nuclear energy was instrumental in getting the United States through its most serious energy crisis in history. During the petroleum shortage of last winter, we estimate that nuclear power saved the equivalent of three billion barrels of oil. In some areas in the North, nuclear power accounted for as much as 20 percent of the electric capacity during that trying period. During this inflationary period, when oil, gas and coal were climbing in price beyond all expectations, nuclear power demonstrated its invaluable cost savings in area after area around the country. Yet the principal trend in the United States' nuclear industry has not been expansion, but deferrals and, in some instances, cancellations.

The widely publicized postponements of nuclear power plants were caused partly by the loss of a year's growth in electricity demand, as a result of energy-savings programs and higher costs. A larger factor, though, seems to have been the financial pressures that hit utilities. Fuel costs and interest charges continued rapid increases, while the availability of capital and utilities' stock value dropped to their lowest points in years. These factors, on top of regulatory requirements that seem punitive in comparison with any other industry's, have served to discourage nuclear power just when it is most needed. As a result many utilities that recognize the advantages of nuclear energy are being forced to build fossil plants, because of their lower capital requirements—and in many cases those plants represent unnecessary consumption of fuel resources and unnecessary fuel costs of up to a half-billion dollars for the utility customers.

So the industry's first two decades seem to be ending in the form of good news/bad news. Good news: Nuclear power has developed into a commercial, economical alternative to fossil fuels just at the time that it is desperately needed. Bad news: Utilities are being discouraged from using it. . . .

. . . I would like to . . . make some specific recommendations for these parties that I think are important if we are to work our way out of this national dilemma.

First, for the federal government, which through its diverse forms is the predominant factor in our energy policies. It is difficult to implement a single, coherent energy policy in a society as diverse as ours, and such a consistent policy does not seem likely in the near future. But as a minimum the government needs to expedite its analysis of our energy problems, recognize the few realistic alternatives available, and get on with the job of encouraging them. Hopefully the studies and the reorganizations are close to an end. We need to be able to plan with a little more confidence that today's policy will also be tomorrow's policy.

This is especially necessary in the nuclear power field, where government must recognize its responsibility. There is probably no private activity as stringently regulated as nuclear power. The mounting regulations and requirements—many in the name of the environment or public disclosure—often effectively penalize nuclear power as a competitive energy source. Other policy decisions, especially in licensing and the fuel cycle, must be expedited if nuclear power is to expand its contribution to our energy supply.

One of the results of the various energy policy studies that have been conducted by government in the past two years should be a realistic estimate of the role that nuclear power can play, and should play, in the decades to come. Such an estimate will be meaningful only if there is a coordinated review of all the factors that go into the nuclear power program—licensing, enrichment, fuel cycle activities, financing and others. I would recommend that a joint government-industry commission be charged with defining a goal for nuclear power in the future—and, of course, other energy forms—and examining the factors that could be obstacles to meeting that goal. A great deal of groundwork for such a comprehensive effort has been done by various segments of government and industry, and it now seems time for an authoritive [*sic*] joint panel to collate and begin implementing the findings. . . .

The industry—utilities and suppliers—must recognize their own responsibility to the public, not merely in meeting next year's demand, but in working toward long-term solutions to our problems. Every factor in the solution, even those that may rub against traditional profit incentives, should be encouraged, including energy conservation, improved efficiency and serious development of additional power sources.

Another part of the industry's responsibility is to find a way to inform the public better about these complexities. Our situation is not an easily understood one, even by those of us in the field. To the public, it must be incomprehensible.

A major factor in finding our way out of this maze must be the cooperation of government and the public, and we cannot expect that until they better understand the realities of our energy situation. It is sadly ironical that much of industry is being forced to cut back its information programs, because of economic pressures, just at the time that we most urgently need to get an important message to the public. Both the utilities and their regulators must recognize and fulfill the responsibility of informing the public fully and fairly.

This leads to the responsibility of a related group, the media. The role of the press in U.S. policy decisions seems to have become more important than ever before, as our society and its policies have become geometrically more complex. In many of these areas the complexities and the sheer importance of the subject seem to have outrun the traditional journalistic methods of dealing with them. The energy story is a case in point. Our alternatives are not simple either/or situations, with industry on one side and the environment on the other, or with a powerful nuclear establishment weighted against other energy sources, such as solar or geothermal, that would be preferable. The issues and the conflicts are much more complicated, more subtle, and ultimately more important. The conventional media approach to stories that smack of "controversy" has been to pit two spokesmen from opposing points of view against each other, as in a boxing match. This, I would contend, usually leads to hard-headed polarization and an oversimplification of the issues, which ultimately misleads the public. On issues of such critical importance, I would hope that the most influential media would find more thoughtful ways to educate the public about the complexities.

I'm sure it will surprise no one to learn that I also have a few suggestions for critics of nuclear power. Everyone seriously concerned about health and safety, and about the environment, must begin realizing that there are no absolutes in life. Distasteful as it may be, we have to think in terms of costs and benefits, of tradeoffs, of comparative effects. As far as I know, every group of experts that has done a comparative study of realistic energy forms has concluded that nuclear power is the least harmful to the environment and at least as safe as any alternative. Demanding that all effects and risks be kept to zero represents an obstructionalism that has contributed more to our problems than it ever will to our solutions.

Everyone seriously concerned about conservation should also recognize that our energy resources themselves are finite. With today's technology, only through the use of the nuclear fuel cycle . . . can we extend them significantly. In addition, we are learning that our supply of capital is as finite—and perhaps as endangered—as any other resource. Like all resources, it should be used efficiently. It is a gross waste of this diminishing resource to squander it on marginal environmental improvements, merely because a vulnerable regulatory system allows such a demand, when it could be used much more effectively in other environmental or health and safety areas.

COMMITTEE ON THE PRESENT DANGER

"COMMON SENSE AND THE COMMON DANGER," 1976

Not everyone agreed that the pursuit of arms control and détente represented the best course for American foreign policy. The Committee for the Present Danger (CPD) originally formed in 1950 to promote the ideas laid out in National Security Council Report 68—namely, a large buildup of U.S. military capacity. The CPD reformed in 1976 in response to the perceived weakening of American foreign policy positions and military capacity in the wake of the Vietnam War, arms control negotiations, and other détente initiatives. How does the CPD—a group made up largely of business leaders and foreign policy experts—characterize the Soviet Union? What is the vision laid out here for the relationship between political and economic freedom? How does this perspective compare to the nuclear consensus of the 1950s and 1960s?

....

I

Our country is in a period of danger, and the danger is increasing. Unless decisive steps are taken to alert the nation, and to change the course of its policy, our economic and military capacity will become inadequate to assure peace with security.

Excerpted from Committee on the Present Danger, "Common Sense and the Common Danger," in *Alerting America: The Papers of the Committee on the Present Danger*, ed. Charles Tyroler II (Washington, D.C.: Pergamon-Brasseys, 1984).

The threats we face are more subtle and indirect than was once the case. As a result, awareness of danger has diminished in the United States, in the democratic countries with which we are naturally and necessarily allied, and in the developing world.

There is still time for effective action to ensure the security and prosperity of the nation in peace, through peaceful deterrence and concerted alliance diplomacy. A conscious effort of political will is needed to restore the strength and coherence of our foreign policy; to revive the solidarity of our alliances; to build constructive relations of cooperation with other nations whose interests parallel our own—and on that sound basis to seek reliable conditions of peace with the Soviet Union, rather than an illusory détente.

Only on such a footing can we and the other democratic industrialized nations, acting together, work with the developing nations to create a just and progressive world economy—the necessary condition of our own prosperity and that of the developing nations and Communist nations as well. In that framework, we shall be better able to promote human rights, and to help deal with the great and emerging problems of food, energy, population, and the environment.

II

The principal threat to our nation, to world peace, and to the cause of human freedom is the Soviet drive for dominance based upon an unparalleled military buildup.

The Soviet Union has not altered its long-held goal of a world dominated from a single center—Moscow. It continues, with notable persistence, to take advantage of every opportunity to expand its political and military influence throughout the world: in Europe; in the Middle East and Africa; in Asia; even in Latin America; in all the seas. . . .

For more than a decade, the Soviet Union has been enlarging and improving both its strategic and its conventional military forces far more rapidly than the United States and its allies. Soviet military power and its rate of growth cannot be explained or justified by considerations of self-defense. The Soviet Union is consciously seeking what its spokesmen call "visible preponderance" for the Soviet sphere. Such preponderance, they explain, will permit the Soviet Union "to transform the conditions of world politics" and determine the direction of its development.

The process of Soviet expansion and the worldwide deployment of its military power threaten our interest in the political independence of our friends and allies, their and our fair access to raw materials, the freedom of the seas, and in avoiding a preponderance of adversary power. . . .

III

Soviet expansionism threatens to destroy the world balance of forces on which the survival of freedom depends. If we see the world as it is, and restore our will, our strength and our self-confidence, we shall find resources and friends enough to counter that threat. There is a crucial moral difference between the two superpowers in their character and objectives. The United States—imperfect as it is—is essential to the hopes of those countries which desire to develop their societies in their own ways, free of coercion.

To sustain an effective foreign policy, economic strength, military strength, and a commitment to leadership are essential. We must restore an allied defense posture capable of deterrence at each significant level and in those theaters vital to our interests. The goal of our strategic forces should be to prevent the use of, or the credible threat to use, strategic weapons in world politics; that of our conventional forces, to prevent other forms of aggression directed against our interests. Without a stable balance of forces in the world and policies of collective defense based upon it, no other objective of our foreign policy is attainable.

As a percentage of Gross National Product, U.S. defense spending is lower than at any time in twenty-five years. For the United States to be free, secure and influential, higher levels of spending are now required for our ready land, sea, and air forces, our strategic deterrent, and, above all, the continuing modernization of those forces through research and development. The increased level of spending required is well within our means so long as we insist on all feasible efficiency in our defense spending. We must also expect our allies to bear their fair share of the burden of defense.

From a strong foundation, we can pursue a positive and confident diplomacy, addressed to the full array of our economic, political and social interests in world politics. It is only on this basis that we can expect successfully to negotiate hardheaded and verifiable agreements to control and reduce armaments.

RALPH W. DEUSTER

"R_x FOR THE 'BACK' OF THE CYCLE," 1976

In the 1970s the storage of radioactive waste remained one of the most perplexing problems for both military and civilian nuclear energy. When engineers designed the first nuclear reactors, they planned for spent fuel to undergo a procedure known as reprocessing. Dissolving spent fuel rods with nitric acid allowed for the chemical separation of plutonium and uranium. Reprocessing made 99 percent of the uranium and plutonium from the spent fuel available for commercial reuse in the form of mixed oxide fuel. Experts regarded the use of mixed oxide fuels and "breeder" reactors that could run on plutonium as necessary for the expansion of the nuclear power industry. But making plutonium available as commercial fuel created an additional worry about the availability of weapons-grade fuel and the specter of proliferation—the spread of nuclear weapons to other nations.

Furthermore, the commercial viability of reprocessing remained unclear. Concerns about waste management and reprocessing threatened the basic economic equations of commercial nuclear power. Ralph W. Deuster worked for Nuclear Fuel Services, a company that provided fuel for both military and civilian reactors. In this speech, delivered at a 1976 conference on the fuel cycle, Deuster explains what he sees as steps necessary to solve the radioactive waste dilemma. What does he "diagnose" as the problems facing the nuclear power industry?

Excerpted from Ralph W. Deuster, "Rx for the 'Back' of the Cycle," Subcommittee on Environment and Safety, Joint Committee on Atomic Energy, *Radioactive Waste Management*, 94th Congress, 2nd session, 1976, 348–52.

How does he characterize opponents of nuclear energy? What does he see as their ulterior motives?

• •

In preparing my own remarks, I've thought about several annoying symptoms in addition to those of the obvious aching "back" of the fuel cycle. . . .

The first ailment is national in scope. It relates to the obvious need to increase the generation of electricity from coal and the atom . . . and the failure of some politicians and part of the public to perceive the facts clearly.

In this bicentennial year, it's appropriate to consider the beginnings of the commercial nuclear age, some 22 years ago, subsequent to the passage of the 1954 Atomic Energy Act. In those days, the shift of emphasis from the destructive to the peaceful atom was considered a most patriotic move. Many people who transferred to the peaceful atom, as I did, were so motivated, as well as by the atom's tremendous energy content, compatibility with a decent environment, and future savings of hydrocarbon fuels for better things than just burning. Many of us recognized, even then, that conventional fuels would eventually be in short supply.

It may seem corny to some, but I still consider myself to be patriotic. I'm still dedicated and convinced of the same advantages for nuclear that I saw when I entered the business. Nuclear does reduce pollution, it does make available almost limitless energy reserves for electrical power generation purposes and it does reduce U.S. dependence on foreign sources of petroleum and gas. To me, these advantages seem like motherhood, they're so obvious. But these advantages seem to have escaped the ken of the public and much of the Congress. So we must continue the fight to win the battle of public support for the atom—not merely for business reasons, but because our country needs it.

We, here today, should press for as much energy independence as the U.S. can achieve by using coal and the atom; and I strongly urge that in this bicentennial year we endorse such a view as a national goal. The fact is that we have no other choices for electrical energy. Suggestions that fusion and solar will solve our needs in the next quarter century are shamefully misleading. Further, without reprocessing and plutonium recycle, there is no breeder program and without the breeder program, we could easily have to give up our standard of living, and our important roles in world affairs. So, it is our duty to help set the facts straight for the public.

I won't pull any punches in describing the virus which I think has clouded our national vision. I believe that some of the most violent anti-nuclear attacks are a subterfuge for other motivations. Some "anti-nuclears" really are motivated by anti-growth and anti-capitalism ideas—and there is plenty of evidence to support this view. They are attacking nuclear because it holds the key to energy growth, and a nuclear success will spoil a no-growth prospect. Yet I can't understand why [Ralph] Nader, [Barry] Commoner and others consider a no-growth doctrine to be a panacea for the people. Shortages of energy, not to mention the increases in cost that these anti-nuclear acts are creating, bring the greatest hardship on the very persons they claim to be championing.

What's my prescription? . . . Well, we must talk about it frankly. For if we fail to label such socio-economic nonsense as nonsense, we're contributing to that nonsense, and potentially helping to deprive our society of the energy it desperately needs. . . .

So now let's look at that "aching back." First, let me try to specify the ailments, then make a diagnosis, and finally write some prescriptions. The first problem is that the "back" is very stiff from lack of exercise. After all, the last time the "back" of the cycle was used was in 1971. Since then the "back" has been in traction! So we must prescribe a way to get existing plants on their feet and running. But the complications for this patient are quite significant. In addition to the lack of exercise and bitter regulatory pills, we will find that once the facilities become productive again, plutonium will be available in greater quantities; and that will start some observers shouting "poison!" The patent medicine press likes to alarm people about plutonium—they use the phrase "the most dangerous material known to man." I say "Phooey!" I certainly don't regard plutonium lightly; but it's worth noting that students of the material, and persons who have worked with it for extended periods, find it very manageable. Nevertheless, plutonium, its recycling, and the security provisions which must be established to prevent its diversion, are ailments of the "back" of the cycle that should be dealt with. The Nuclear Regulatory Commission is dealing with those problems on a quote "accelerated schedule" and, I think, responsibly. . . .

Another reason for that aching "back" is constipation. The first blockage to give us trouble is the delay in reprocessing spent fuel. Another blockage is that there is still no designated spot to dispose of the high level waste, and the anti-nuclear forces have really made a big effort to publicize that problem! In these blockage cases, the treatment involves a number of steps and surely a laxative. There is a need:

1. For formal classification of wastes based on realistic risk/benefit analyses (and that includes realistic, logical definitions for high level, intermediate, low level and transuranic wastes), or by whatever classification;
2. To define processes in waste treatment;
3. To identify packaging criteria and disposal choices for the various waste classifications;
4. For consistent, unambiguous federal regulations and guides;
5. To establish the government repositories to accept the various waste classes;
6. To develop cost projections for permanent disposal;

All of these on a schedule that is consistent with industry's needs.

I don't mean to suggest that the cure is up to government alone—far from it! Industry has the responsibility to pursue the waste processing and packaging activities, to operate such facilities, and to provide technical support and knowhow to operate the state-licensed low-level waste burial grounds. But ERDA [the Energy Research and Development Administration] has the responsibility to move forward rapidly with the establishment of waste classifications and waste packaging criteria, and NRC [the Nuclear Regulatory Commission] has the responsibility to issue regulations that conform with the National Environmental Protection Act.

Although ERDA has proposed a substantial and elaborate Waste Management Program for Fiscal 1977, I, personally, am dissatisfied with the progress that is being made toward its implementation. We face a far longer wait than necessary before something substantive is defined. The current plan of publishing a document that lists all of the alternatives, then having a conference to discuss these alternatives, all as preparatory to the rewrite of a new Draft Environmental Statement on Waste, is certainly not the most dynamic route to completion.

LEONARD RIFAS

ALL-ATOMIC COMICS, 1976

In 1976 cartoonist Leonard Rifas joined a growing chorus of antinuclear activists. This statement appeared inside the front cover of *All-Atomic Comics*, in which Rifas laid out the case against nuclear power. In what different ways does a comic book convey social and environmental messages? How does Rifas echo the messages suggested by other antinuclear activists in questioning scientific expertise, decision-making authority, and the obligations of citizenship? What contributed to his diminished faith in scientific authority?

....

This is the third printing of a comic book I drew in 1976 about nuclear power and its dangers. At that time the promoters of nuclear power were telling us that we needed nuclear power for cheap energy and to create jobs. Since then, further studies have shown that nuclear power is *more* expensive than its alternatives and that it *destroys* more jobs than it creates.

When I first started criticizing nuclear power I felt uncomfortable to be going against "the experts" who said that nuclear power was safe, clean, and cheap. I unconsciously interpreted the fact that I often couldn't understand what they were talking about as a sign of their superior intelligence.

I realize now the illusion of imagining that energy policy decisions of national importance can be made fairly or equitably by the specialists

---

Excerpted from Leonard Rifas, *All-Atomic Comics* (San Francisco: Educomics, 1976). Courtesy of Leonard Rifas.

hired by the energy corporations and their counterparts in government. I understand better now how political pressures influence what studies get funded and which experts are chosen to conduct them. . . .

The interests of the energy corporations and the interests of society are not the same. In fact, we now have an increasingly detailed idea of an alternative energy future which would give us a cheaper, safer, more reliable, more job-creating energy system than the one the energy corporations have planned for us.

Who can we trust to struggle for our best interests? Ourselves, that's who. Our busy, underinformed selves, and the sooner we get working on it the better.

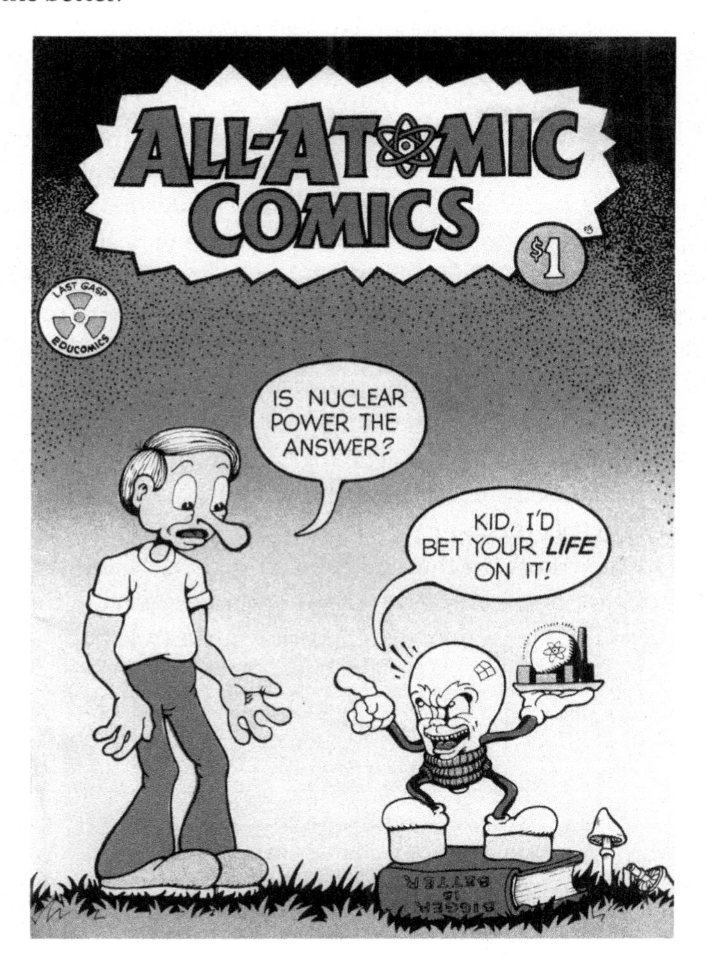

DAVID N. MERRILL

NUCLEAR SITING AND LICENSING PROCESS, 1978

In the late 1970s, Seabrook Station Nuclear Power Plant in New Hampshire became the site of the largest and most controversial antinuclear demonstrations in the nation. A protest group calling itself the Clamshell Alliance formed in opposition to the plant, pioneering direct-action techniques to slow down construction and raise awareness about nuclear power. In one 1977 rally, over fourteen hundred protesters were arrested for trespassing on the construction site. By the time Seabrook Station came online in 1990, the Public Service Company of New Hampshire—the largest single owner of the plant—had filed for bankruptcy, largely because of debt acquired in financing the plant's construction and the long delay in its ability to generate power and revenue from the project. Utility executive David Merrill testified at hearings called by Congress in 1978 to investigate the protests and the licensing process for Seabrook Station. What does he see as the problem facing the nuclear industry, and how does he expect the industry to fare in the future? How do his concerns compare to those of other stakeholders in the debates over nuclear power, and how does he engage ongoing conversations about the proper role of government as a regulatory authority for the industry?

. . . .

Excerpted from David N. Merrill, Testimony, Subcommittee on Energy and the Environment, House Committee on Interior and Insular Affairs, *Nuclear Siting and Licensing Process (Seabrook Station, N.H.)*, 95th Congress, 2nd session, 1978, 157–58.

The licensing process has been variously called, the "Seabrook Saga," "Snarled Seabrook," "the Seabrook Tragedy," "a paradigm of fragmented and uncoordinated government decision making."

And in the privacy of our office, it has been called some other colorful phrases. Regardless of what it has been called, it started on February 1, 1972, with an application to the Site Evaluation Committee (SEC) and the Public Utilities Commission of the State of New Hampshire for a Certificate of Site and Facility. While that action took place an unbelievable six and a half years ago, we actually started developing the Seabrook project ten years ago in 1968. And it will be at least six more years before it is completed.

In the two years after the 1972 filing, the SEC, which is made up of persons from eleven agencies of state government, after a discovery process held thirty-two days of an adjudicatory type of hearing which produced 5,870 pages of testimony plus 107 Company exhibits, plus nearly as many intervenor exhibits totaling several thousand pages. During those hearings, a wide variety of topics were investigated, need for power, alternate sources, alternative sites, water, land and air ecology, transmission line routing, financial ramifications, seismic concerns, aesthetics, nuclear safety, decommissioning, low level radiation and you name it. These subjects are certainly all valid subjects for review. However, before we ever received the Certificate of Site and Facility from the State of New Hampshire, we had filed an application with the NRC [Nuclear Regulatory Commission] and for the next twenty-six months off and on, we went over the same subjects again plus some new ones, and to this day either in the agencies or the courts, six and one-half years after the start of the process, we are still arguing the various points.

There must be a better way. From the utility standpoint, I think we are in danger of losing the nuclear option. The uncertainties in the present nuclear licensing process reflect detrimentally into the daily conduct of the rest of our business to the point that the industry will not willingly proceed with the development of nuclear power even though they may recognize it as the best alternative. . . .

Either the development of nuclear power is the right course of action for the country or it isn't. We cannot continue in the present purgatory—immeshed [*sic*] in various shades of grey.

HELEN CALDICOTT

NUCLEAR MADNESS, 1978

By the late 1970s the antinuclear movement was in full swing. Opponents of nuclear power raised questions about the safety of commercial reactors, the perils of radioactive waste, and a host of other issues. Australian-born pediatrician Helen Caldicott emerged as one of the most prominent national voices in the antinuclear movement. At the time that she published the book excerpted here, Caldicott taught pediatrics at the Harvard Medical School. In what ways does she invoke (or question) medical and scientific expertise to make her point about nuclear energy? Who does she want to take action—and why? The antinuclear movement developed during a turbulent time in American history, and it connected to such issues as faith in government, uncertainties about U.S. foreign policy, and concerns about environmental health. In what ways are these concerns reflected in this document?

I am a child of the Atomic Age. I was six years old when American atomic bombs were deployed against the Japanese, and I have grown up with the fear of imminent annihilation by nuclear holocaust. Over the years my fear has increased. Thousands of nuclear weapons are now being built each year. At this writing, 360 nuclear reactors are in operation in thirty countries around the world, 65 of them in the United

---

Excerpted from *Nuclear Madness: What You Can Do*, by Helen Caldicott (New York: W. W. Norton and Company, 1978).

States. Hundreds more are projected by the end of the century. The nuclear facilities stand to inherit the earth.

As a physician, I contend that nuclear technology threatens life on our planet with extinction. If present trends continue, the air we breathe, the food we eat, and the water we drink will soon be contaminated with enough radioactive pollutants to pose a potential health hazard far greater than any plague humanity has ever experienced. Unknowingly exposed to these radioactive poisons, some of us may be developing cancer right now. Others may be passing damaged genes, the basic chemical units which transmit hereditary characteristics, to future generations. And more of us will inevitably be affected unless we bring about a drastic reversal of our government's pronuclear policies. . . .

Most Americans with whom I've spoken know very little about the medical hazards posed by nuclear radiation and seem to have forgotten (or suppressed) the atomic-bomb anxiety that was so prevalent in the 1950s. Back then, most Americans were aware of the devastation that nuclear war implies: Children were constantly diving under their desks and crouching in school corridors during simulated nuclear attacks while the bomb shelter business boomed and the nation prepared a program of civil defense. Americans were justifiably scared. People recognized that nuclear disarmament could mean the difference between a secure future for them and their children, and no future at all.

But during the 1960s the American public became preoccupied with other matters: political assassination, the Civil Rights movement, the Vietnam war. The threat of nuclear holocaust was submerged by these more immediate problems. Only the Pentagon sustained its interest in nuclear development; in the effort to keep ahead of the Soviet Union, America's military strategists continued to stockpile more and more atomic weapons, while developing and refining delivery systems, satellites, and submarines.

Today, however, it is of the utmost urgency that we refocus our attention on the problems posed by nuclear technology, for we have entered and are rapidly passing through a new phase of the Atomic Age. Despite the fact that reactor technology is beset with hazardous shortcomings that threaten the health and well-being of the nations that employ it, nuclear power plants are spreading throughout the world. Moreover, by making "peaceful" nuclear technology available to any nation wealthy enough to buy a nuclear reactor, we are inviting other countries to join the international "nuclear club" militarily, as well as economically. Following India's lead in 1974, many nations are bound to explode test devices in the coming years. Because of this proliferation of nuclear weapons, the likelihood of nuclear war, the most ominous threat to public health imaginable, becomes greater every day.

In view of the threat that nuclear technology poses to the ecosphere, we must acknowledge that Homo sapiens has reached an evolutionary turning point. Thousands of tons of radioactive materials, released by nuclear explosions and reactor spills, are now dispersing through the environment. Nonbiodegradable, and some potent virtually forever, these toxic nuclear materials will continue to accumulate and eventually their effects on the biosphere and on human beings will be grave: many people will begin to develop and die of cancer; or their reproductive genes will mutate, resulting in an increased incidence of congenitally deformed and diseased offspring—not just in the next generation, but for the rest of time. An all-out nuclear war would kill millions of people and accelerate these biological hazards among the survivors: the earth would be poisoned and laid waste, rendered uninhabitable for aeons. . . .

My experience with national political leaders has not inspired me with confidence. I find that they are generally ignorant of the major medical and scientific ramifications of their decisions. They are manipulated by powerful, well-financed industrial and military lobbies. Driven by power and the need for ego-gratification, they are, to a large degree, desensitized to reality. Their vision is limited to their meager two-to-four-year terms of office; the desire for reelection influences all their decisions. I have found, sadly, that a global view of reality and a sense of moral responsibility for humanity's future are very rare among political figures.

Even after Watergate, Vietnam, and CIA exposes, the American people still seem to trust their political leaders. When I speak at church meetings, describing the enormity of our nuclear madness, people approach me and ask, "The politicians don't know this, do they? Because they wouldn't let any of this happen if they did." I look them in the eye and tell them that their government is totally responsible for organizing this calamity.

But if we can't trust our representatives, whom can we trust? The answer is simple: No one but ourselves. We must educate ourselves about the medical, scientific and military realities, and then move powerfully as individuals accepting full responsibility for preserving our planet for our descendants. Using all our initiative and creativity, we must struggle to convert our democratic system into a society working for life rather than death. . . .

America's scientific community has a unique obligation to assess the morality of its work and to assist the public to understand the perils we face; in the event of nuclear holocaust, it is science that will have led us down the road to self-destruction. The first meltdown or nuclear-waste explosion will disprove, once and for all, the blithe assurances offered by industry experts regarding reactor safety and waste-disposal technology, but only after bringing tragedy into the lives of thousands of human beings.

The Industrial Age has enthroned science as its new religion; the scientific establishment has, in turn, promised to cure disease, prolong life, and master the environment. Thus, humanity now believes that it owns the earth; we have forgotten that we belong to it, and that if we do not obey the natural laws of life and survival, we will all cease to exist.

Unfortunately, my experience has taught me that we cannot rely upon our scientists to save us. For one thing, they do not preoccupy themselves with questions of morality. For another, science has become so specialized that even the best scientists are immersed in very narrow areas of research: most lack the time to view the broad results of their endeavor, let alone the willingness to accept responsibility for the awesome destructive capabilities science has developed. When confronted with the realities of our nuclear insanity, they are often either embarrassed or display a cynical fatalism, saying that humanity was not meant to survive anyway. Since about half of America's research scientists and engineers are employed by the military and related industries, they suffer from a profound conflict of interest: many simply prefer not to ponder the consequences of their work too deeply; after all, they have to feed and educate their children—a strange reaction, considering that, as a direct result of their research, their children may not live out their normal life span.

ABALONE ALLIANCE

"DECLARATION OF NUCLEAR RESISTANCE," 1978

In 1965 the utility company Pacific Gas & Electric announced plans to construct a multireactor generating station at Diablo Canyon, near San Luis Obispo, California. The company planned to spend $162 million and have the plant operational in 1972. As safety hearings and site preparation began, a strong antinuclear movement, known as the Abalone Alliance, coalesced in opposition to the proposal. The Abalone Alliance used a variety of nonviolent, direct-action techniques to oppose the plant and slow its construction, including marches, sit-ins, and blockades. For example, nearly five thousand protesters blockaded the gates of the facility in 1978 to prevent access by construction workers; nearly five hundred people were arrested. The Diablo Canyon Plant went online in 1985, with a total construction cost of over $1 billion. The group's "Declaration of Nuclear Resistance" is below. On what grounds did the Abalone Alliance oppose the Diablo Canyon plant? How do the motivations, goals, and tactics of this group differ from other antinuclear statements of previous years?

• •

We are committed to a permanent halt to the construction and operation of nuclear power plants in California. Nuclear power is dangerous to all life. We encourage the real alternatives of conservation and safe, clean, and renewable sources of energy.

Abalone Alliance, "Declaration of Nuclear Resistance," February 26, 1978, www.reclaimingquarterly.org/web/handbook/DA-Handbk-Diablo81-lo.pdf.

To achieve these goals, we join together from throughout the state to form the Abalone Alliance to oppose nuclear power through nonviolent direct action and education.

Beginning with the Diablo Canyon nuclear power plant, our nonviolent action will be directed to all existing and planned nuclear plants in California. We will continue until nuclear power has been completely replaced by a sane and life-affirming energy policy.

We recognize that:

1. The much advertised need for nuclear energy is derived from faulty and inflated projections of consumption based on a profit system hostile to conservation. The United States has 6% of the world's population consuming over 30% of its energy resources. With a rational energy policy and appropriate changes in construction, conservation, and recycling procedures, the alleged "need" for nuclear energy disappears.
2. Nuclear plants are an economic catastrophe. They are unreliable and inefficient. Nuclear power is an extremely capital-intensive technology. In contrast, conservation and solar-related energy technologies will create many more jobs, both permanent and safe, than the atomic industry could ever provide.
3. The centralized nature of nuclear power takes control of energy away from local communities.
4. There is a direct relationship between nuclear power plants and nuclear weapons. The export of nuclear reactors makes possible the spread of nuclear bombs to nations all over the world. The theft of nuclear materials and the sabotage of nuclear facilities pose further threats to our lives and civil liberties.
5. The dangers of nuclear power are intolerable. They range from a continuous flow of low-level radiation which can cause cancer and genetic damage, to the creation of deadly radioactive wastes which must be completely isolated from the environment for 250,000 years, to the destruction of our rivers, lakes and oceans by radioactive and thermal pollution, to the possibility of a major meltdown catastrophe. No material gain, real or imagined, is worth the assault on life itself that nuclear energy represents.

We therefore insist:

1. That not one more cent be spent on nuclear power reactors, except for efforts to dispose of those wastes already created and to decommission those plants now operating.
2. That American energy policy be focused on conservation and the development of solar, wind, tidal, biomass conversion, and other forms of clean and renewable energy in concert with the efficient recycling and fair distribution of energy.
3. That all people who lose jobs through the cancellation of nuclear construction or operation be retrained immediately for jobs in the natural energy field or in other areas.
4. That we end production, testing, stockpiling, and use of nuclear weapons.

We have full confidence that when the true dangers and expense of nuclear energy are made known to the American people, our nation will reject this tragic experiment which has already caused so much loss in economic and material resources, health, environmental quality, and the control over our own lives.

We pledge we will wage a nonviolent direct action campaign:

1. To stop construction and operation of all nuclear plants in California.
2. To promote the realistic alternatives of safe, clean, and renewable sources of energy.
3. To encourage responsible community control of energy production and use.
4. To support efforts to eliminate nuclear weapons.
5. To build a more loving and responsible world for ourselves, our children, and future generations of all living things on this planet.

In our work, we will maintain a discipline of active nonviolence and full respect for all persons we encounter. We will speak and act truthfully and openly, and we will honestly weigh concerns brought to us.

We pledge our solidarity with all other nonviolent efforts to stop nuclear power worldwide.

We appeal to all people to join us.

REPORT OF THE PRESIDENT'S COMMISSION ON THE ACCIDENT AT THREE MILE ISLAND, 1979

On March 28, 1979, one of the two reactors at the Three Mile Island Nuclear Generating Station (TMI) outside Harrisburg, Pennsylvania, suffered a partial meltdown. The amount of radiation released, and the danger to the public, was minimal. The long-term legacy of TMI lies in the confused governmental response to the accident and its impact on public opinion. The governor's office first announced that everything at the plant was under control but later urged pregnant women and children within five miles of the plant to evacuate the region. Public anxiety about the risk of radiation exposure spiked. In the aftermath of the accident, President Jimmy Carter appointed a commission, chaired by Dartmouth College president John G. Kemeny, to investigate the causes and consequences of the accident. What did the Kemeny Commission identify as the cause of the accident at TMI, and what did it see as the incident's lasting significance? How does the commission's report, excerpted below, intersect with the concerns about decision-making, the free flow of information, and the role of the government raised by other stakeholders in the debates over nuclear power?

Our findings do not, standing alone, require the conclusion that nuclear power is inherently too dangerous to permit it to continue and expand as a form of power generation. Neither do they suggest that the nation

---

Excerpted from President's Commission on the Accident at Three Mile Island, *Report of the President's Commission on the Accident at Three Mile Island: The Need for Change, the Legacy of TMI* (Washington, D.C., 1979), 7–9, 18–19.

should move forward aggressively to develop additional commercial nuclear power. They simply state that if the country wishes, for larger reasons, to confront the risks that are inherently associated with nuclear power, fundamental changes are necessary if those risks are to be kept within tolerable limits. . . .

Popular discussions of nuclear power plants tend to concentrate on questions of equipment safety. Equipment can and should be improved to add further safety to nuclear power plants, and some of our recommendations deal with this subject. But as the evidence accumulated, it became clear that the fundamental problems are people-related problems and not equipment problems.

When we say that the basic problems are people-related, we do not mean to limit this term to shortcomings of individual human beings—although those do exist. We mean more generally that our investigation has revealed problems with the "system" that manufactures, operates, and regulates nuclear power plants. There are structural problems in the various organizations, there are deficiencies in various processes, and there is a lack of communication among key individuals and groups.

We are convinced that if the only problems were equipment problems, this Presidential Commission would never have been created. The equipment was sufficiently good that, except for human failures, the major accident at Three Mile Island would have been a minor incident. But, wherever we looked, we found problems with the human beings who operate the plant, with the management that runs the key organization, and with the agency that is charged with assuring the safety of nuclear power plants.

In the testimony we received, one word occurred over and over again. That word is "mindset." At one of our public hearings, Roger Mattson, director of NRC's Division of Systems Safety, used that word five times within a span of 10 minutes. For example: "I think [the] mindset [was] that the operator was a force for good, that if you discounted him, it was a measure of conservatism." In other words, they concentrated on equipment, assuming that the presence of operators could only improve the situation—they would not be part of the problem.

After many years of operation of nuclear power plants, with no evidence that any member of the general public has been hurt, the belief that nuclear power plants are sufficiently safe grew into a conviction.

One must recognize this to understand why many key steps that could have prevented the accident at Three Mile Island were not taken. The Commission is convinced that this attitude must be changed to one that says nuclear power is by its very nature potentially dangerous, and, therefore, one must continually question whether the safeguards already in place are sufficient to prevent major accidents. A comprehensive system is required in which equipment and human beings are treated with equal importance.

We note a preoccupation with regulations. It is, of course, the responsibility of the Nuclear Regulatory Commission to issue regulations to assure the safety of nuclear power plants. However, we are convinced that regulations alone cannot assure safety. Indeed, once regulations become as voluminous and complex as those regulations now in place, they can serve as a negative factor in nuclear safety. The regulations are so complex that immense efforts are required by the utility, by its suppliers, and by the NRC to assure that regulations are complied with. The satisfaction of regulatory requirements is equated with safety. This Commission believes that it is an absorbing concern with safety that will bring about safety—not just the meeting of narrowly prescribed and complex regulations. . . .

The President asked us to investigate whether the public's right to information during the emergency was well served. Our conclusion is again in the negative. However, here there were many different causes, and it is both harder to assign proper responsibility and more difficult to come up with appropriate recommendations. There were serious problems with the sources of information, with how this information was conveyed to the press, and also with the way the press reported what it heard.

We do not find that there was a systematic attempt at a "cover-up" by the sources of information. Some of the official news sources were themselves confused about the facts and there were major disagreements among officials. On the first day of the accident, there was an attempt by the utility to minimize its significance, in spite of substantial evidence that it was serious. Later that week, NRC was the source of exaggerated stories. Due to misinformation, and in one case . . . through the commission of scientific errors, official sources would make statements about radiation already released (or about the imminent likelihood of releases of major amounts of radiation) that were not justified by the facts—at least not if the facts had been correctly understood. And NRC was slow in confirming good news about the hydrogen bubble. On the other hand, the estimated extent of the damage to the core was not fully revealed to the public. . . .

We therefore conclude that, while the extent of the coverage was justified, a combination of confusion and weakness in the sources of information and lack of understanding on the part of the media resulted in the public being poorly served. . . .

We have stated that fundamental changes must occur in organizations, procedures, and, above all, in the attitudes of people. No amount of technical "fixes" will cure this underlying problem. There have been many previous recommendations for greater safety for nuclear power plants, which have had limited impact. What we consider crucial is whether the proposed improvements are carried out by the same organizations (unchanged), with the same kinds of practices and the same attitudes that were prevalent prior to the accident. As long as proposed improvements are carried out in a "business as usual" atmosphere, the fundamental changes necessitated by the accident at Three Mile Island cannot be realized. . . .

Nevertheless, we feel that our findings and recommendations are of vital importance for the future of nuclear power. We are convinced that, unless portions of the industry and its regulatory agency undergo fundamental changes, they will over time totally destroy public confidence and, hence, *they* will be responsible for the elimination of nuclear power as a viable source of energy.

GLORIA GREGERSON

RADIATION EXPOSURE AND COMPENSATION, 1981

The nuclear consensus generated long-term social and environmental consequences. This was particularly, and painfully, clear as thousands of people suffered the impacts of radiation exposure. In 1981 the U.S. Senate held hearings on the prospect of compensating three groups of people for this exposure: the "downwinders" who lived in the fallout path of the explosions detonated at the Nevada Test Site; military veterans whose service required their deliberate exposure to radiation during testing both in Nevada and in the Pacific; and uranium miners and millworkers. Congress did not pass the Radiation Exposure Compensation Act until 1990.

In the testimony excerpted below, downwinder Gloria Gregerson recalls the experiences of growing up near the Nevada Test Site and the health problems that she experienced in subsequent years. What does she remember about the way that members of her community were informed about the tests and the risks they posed? What attitudes does she display toward scientists, government, and protesters? Who does she blame for her troubles? Why and how does she invoke her political conservatism—and how might this have colored her reactions to nuclear testing—in the 1950s as well as in 1981?

Excerpted from Gloria Gregerson, Senate Committee on Labor and Human Resources, *Radiation Exposure Compensation Act of 1981*, 97th Congress, 1st session, 1981, 146–52.

I was living downwind in Bunkerville, Nevada during all the years of the atomic testing. The first blast came without any warning. No one was informed it was going to happen. The flash was so bright, it awakened us out of a sound sleep. We lived in an old two story home, and when the blast hit, it not only broke out several windows, but it made two large cracks full length of the house. One on the North and one on the South.

After the first blast, my parents wouldn't let us stay in the house. They would load all of us, still in our pajamas, in the car and drive to the top of a nearby hill. . . .

The Government officials came to our school to talk in an assembly on several occasions. This was only after several shots had already been shot off. They would always preceed [*sic*] their comments with the saying "there's nothing to be alarmed about, nothing to worry about but—"

Some of the things they would say to us were:

1. Wash your car everyday.
2. Wash your clothes at least twice before you wear them.
3. Spray water on the trees, lawns, plants and vegetation before touching or walking on them.
4. Don't drink the local milk, yet at that time there wasn't access to any other kind.
5. Don't worry about anything, there is nothing to harm you.

This they kept emphasizing, but why take the trouble to come all that way and take time to hold an assembly just to tell us there was nothing to worry about?

We were given badges to wear and we were monitered [*sic*] numerous times. We were never told the results of those readings though.

I think one of the most important factors is that we were never warned about what could happen if we received any radiation.

As a young girl, I remember playing around and under the trees shaking the white powdered dust all over me. I thought if [*sic*] was fun. I also remember writing my name in the dust all over the cars on numerous occasions. . . .

When I was 16 years old, I found I had cancer in my female organs.

After numerous operations to remove the cancer, I finally had to have a hysterectomy 2 years after I graduated from high school. I have never been able to have children. In my late 20's early 30's, I had numerous operations for another type of cancer, squamous cell carsinomas [*sic*].

I have adopted 5 children. 4 all on one day and in October 1978 I received a one month old baby boy. Three months later, I wasn't able to care for my family and was hospitalized in January only to find that my blood was so low the doctor said I probably wouldn't have survived the day without the transfusions I received. . . . My life expectancy was 3 weeks.

My husband and sister and mother raised my baby for the next 4 months while I was confined to the hospital. He was over 7 months old before I got to hold him and take care of him a small amount of the time. During this time he had forgotten me and thought my sister Judy was his mother. This kind of situation is heartbreaking to a mother, especially when she knows that this baby is the only one she will ever have.

All of my children have suffered so much and are still suffering not knowing if they are going to be motherless again, not to mention how it upsets me knowing I could die in a few months or even weeks. . . .

In my opinion, the Government's attitude on the subject of the fallout victims and atomic testing is shameful. The pain, horror and suffering brought to us, the innocent victims, and to our families is so great, and yet the Government feels we have no justified complaint and we are looked upon by some as illiterate and fortune hunters in our suit to receive justified compensation for our medical bills and in our attempts to stop the continued atomic testing. . . .

I have a few questions to ask that I hope everyone will ponder.

1. What gives the Government the right to experiment with my health and the health of my children and my children's children?
2. Who is it in the Government that is responsible for further testing? What type of cold blooded man can be in charge of deliberately purpotrating the radio-active atrosity that is still taking place upon fellow americans [*sic*]?
3. If the Government has spent $175,000,000.00 studying Nagasaki and Hiroshima, why are they so reluctant to help and study the fallout victims of people in our own nation?

I would like the ones responsible for the testing to come to me and tell me face to face that radio-activity fallout did not cause my leukemia.

Please keep in mind before I make these next few statements that I am not anti-science or anti-intellectual. I was the Salutatorian of my class in high school. My husband has doctorate degree. We resent the attitude some scientists and doctors take that we can't understand anything scientific. I issue a challenge to the scientific community in general. Who are you to say that low-level radiation won't effect [*sic*] my children's genes? Who are you to experiment with my life?

One thing that bothers us is the quick smug reassuring answers that radiation couldn't have caused these problems. If you are so sure of your answers, why not let the clouds blow over Los Angeles?

Has our scientific community reached perfection? Have we now learned all the answers? Does the medical profession now have all the answers regarding human health and disease?

I don't intend to spend what little time I have left in a vindictive "Ban the Bomb" exercise, but I can't say the same for my children and husband.

Some look upon the Government as a nameless body that can't be held responsible, but the Government is responsible. It is made up of individuals that we the people have elected. I want to know the names of those responsible for the testing and the names of those responsible for the testing in the 1950's. They are people and the companies that are responsible.

I want to know who this nameless person is that was elected by the people and is being paid for testing so that when I die, my baby can go to him and ask him what he learned from the tests, and then let my sons and daughters decide if it was worth the life of their mother.

What gives them the right to do this to experiment with my health and the health of my babies? Who is it that doesn't have to answer to anyone about this?

I always thought that people that ranted against the Government Bureaucrats were a little crazy and radical, and now I find myself asking the same questions. I am not a so called bleeding heart liberal[.] I am and always have been very conservative. Who are the bureaucrats who gave them the right. Who's job are they protecting? If the testings stopped, they would lose their jobs.

PART 5

RENEWAL

American ideas about nuclear energy realigned in the early 1980s. After a nearly two-decade lull, nuclear fear flared, fueled by the election of Ronald Reagan in 1980 and his renewal of Cold War rhetoric. Reagan's aggressive posturing increased tensions, and his support for defensive nuclear technologies like the Strategic Defense Initiative threatened to destabilize the precarious nuclear balance between the United States and the Soviet Union. The peace movement experienced a resurgence in the 1980s as well, galvanized by predictions of environmental destruction and human extinction that would surely follow a nuclear war. Meanwhile, the commercial nuclear power industry remained mired in a broken regulatory system and hampered by public distrust in the wake of the accident at Three Mile Island. And yet many experts predicted a resurgence for the industry in the 1980s—a development sure to be aided by President Reagan's commitment to enhancing individual and economic freedom by limiting the government's regulatory authority. These patterns shifted again, especially in terms of nuclear weapons, at the end of the decade, when the United States and the U.S.S.R. agreed on a series of arms control treaties that signaled the end of the Cold War.

The documents in part 5 illustrate these changing patterns of nuclear thought. How does the vision of progress, prosperity, and freedom laid out by Reagan compare to the nuclear consensus of the 1950s? How do the challenges to this renewed consensus compare with the challenges of previous decades? How did ideas about nature, civil society, and progress change as the Cold War came to its surprising end?

DAVID E. LILIENTHAL

ATOMIC ENERGY: A NEW START, 1980

Over the course of his long and distinguished career as a government administrator, writer, and businessman, David Lilienthal offered several in-depth commentaries on nuclear issues. In the 1960s he argued that the nuclear industry should not and could not develop until it solved the problem of radioactive waste, making him among the first commentators to identify waste as a crucial issue for commercial nuclear power. In the aftermath of the accident at Three Mile Island and with the commercial nuclear power industry at a standstill, Lilienthal nevertheless envisioned a bright future for nuclear power. What does he hope it can achieve, and why does he think nuclear energy is necessary? What commentary does he make on the values of the nuclear consensus?

Nuclear energy is by no means finished; it remains one of the great hopes of mankind, and in due course it will play a major role, perhaps the decisive role in providing the energy that the world needs so badly. But that goal will not be reached on the road we are now traveling. We need to back away from our present nuclear state in order to find a better way, a route less hazardous to human health and to the peace of the world and its very survival.

The decade of the 1980s will be crucial not only for the future of atomic power but for the broader issues that have been summed up

---

and illuminated by the dramatic worldwide nuclear controversy. Never before have we been so gravely menaced by what our science and technology have created—and what the powers and principalities of the earth have so signally failed to put to use and control in a decent and humane way.

The citizen protest against atomic energy plants here and abroad was not raised against nuclear hazards alone. Nor would it be satisfied if all nuclear plants were to be closed tomorrow. To a large extent it has been a protest against the misuse of science, the misdirection of enormous forces that human ingenuity has brought into being. It is a protest against the abuses of industrial technology that poison the land instead of nurturing it, that sour the air and foul the water, that devour marsh and woodland and make hazardous to health and peace of mind the cities and factories in which people live and work. It is a protest against governments—all governments—that spend billions in an endless, insane atomic arms race that consumes the cream of the world's resources and much of its brightest human talent. It is a protest against the uses of science and technology that are antihuman and antilife.

This is not to say that today's protesting citizen groups do not include among them individuals who are arrogant, ignorant, and self-seeking. Many of the demands and arguments that they put forward are plainly exaggerated, even hypocritical.

Nevertheless there is a strong, genuine, largely spontaneous surge of feeling that animates these protests, a feeling that expresses, in crude ways at times, the anxieties of a great part of the public. I sense that people in general know that something has gone wrong not just with nuclear power but with the industrial and technological forces that produced it in its present hazardous form. These misgivings are part of a larger sense of worry and unease about our society itself. Ordinary citizens find their well-being threatened, their serenity shattered; they have doubts about the food they eat, the air they breathe. They are aware of a growing frustration, a feeling that it is time to call a halt. Where better to call a halt than with the most mysterious and dramatically menacing of all forces, the atom? . . .

Now, energy for what and for whom?

No American energy policy makes sense unless it takes into account the hopes and needs of *all* peoples. Our international political position has been damaged by the assertions of our own introspective energy "experts" who have branded us heedless gobblers of resources, particularly energy resources, when it is these very energy resources that have enabled us to aid the helpless the world over.

"Conservation" is advanced passionately as the answer to America's energy problem. If conservation means minimizing obvious waste and increasing efficiency, how can anyone possibly oppose such a self-evident proposition? But if conservation is intended as the principal means of achieving American energy self-sufficiency—as some assert—isn't this nothing more than a kind of isolationism in a particularly heartless form, an elitist disguise to mask putting a limit on total energy production thereby slowing economic growth for those who need growth the most? In urging a conservation policy of "making do with less and less," we need to remember that most of the world's people have little or no energy to "conserve."

We grumble when there is a temporary shortage of gasoline for our cars, when a brownout briefly stalls elevators and melts the ice cubes in the freezer. But the overwhelming majority of human beings have no gasoline, no cars, no elevators, no refrigerators. Their countries have never had much energy, they do not have much today, and they will not have much more in the foreseeable future.

It requires little imagination to picture what goes on in the mind of an individual from one of these countries who visits the United States and overhears our talk of an energy crisis while observing our cities lighted up like giant jukeboxes, our highways choked with cars and trucks, and all the other evidence of our powerhouse civilization.

Our future energy policy may be a technical success, but it will be a moral and political failure if it aims only at filling our gas tanks and keeping our thermostat settings high. Such a limited, nationalistic energy policy will inevitably pit us, the richest of the rich, against the less developed nations in the greedy business of grabbing everything we can for ourselves. And because there just isn't enough to go around, and there probably never will be enough, our ill-tempered reactions to minor energy shortages will worsen. We then run the risk of yielding to the recurrent proposals to ready the Marines for the seizure of the oil fields of the Persian Gulf states.

The only escape from this sinister and destructive meanness is to frame policies that recognize our ethical responsibility toward the other peoples on this planet as fellow members of a world community. This task cannot be dismissed as empty idealism; it is the only practical policy for the long run. Unless we make a major contribution toward easing the *world's* energy shortage—instead of merely satisfying our own needs—we may be creating for our children's children a life of constant crisis and chronic insecurity. . . .

The American dream of plenty has come true.

Yet there is more to that dream than plenty. There must be, for otherwise we are no more than what our detractors say we are, an overfed, self-indulgent nation of consumers. I do not believe this is true. Every day we see strong evidence that we are as capable now as we have been in the past of seeing beyond our own needs, of looking with understanding and acting with fairness toward other peoples.

This I do believe: our energy policy will be a test of our moral worth and our place in history. It will be a sign and a testament; it will show the true face of America to the world; it will tell us, too, what kind of people we really are.

With the other nations of the world we share a common problem, the necessity for finding the energy that today and in the years ahead we all need so urgently. . . . Now, belatedly, we recognize that our nuclear technology is not really so advanced. It isn't dependable enough, it isn't safe enough. And we have the responsibility to ourselves and to other peoples and nations to begin at once to create a better answer, no matter how long it takes. When we have it—and we *will* have it some day—we will have satisfied a historic obligation as well as a desperate necessity. It would be most appropriate if the nation that made and used the first atomic bomb should be the nation that creates the first safe atomic power.

RONALD REAGAN

"ADDRESS TO MEMBERS OF THE BRITISH PARLIAMENT," 1982

The election of President Ronald Reagan in 1980 signaled a sharp change in both domestic and foreign policy. At home, Reagan promised that a smaller federal government and a deregulated economy would return the country to the prosperity it had lost in the 1970s. Abroad, he ended American commitment to détente and renewed the rhetoric of the Cold War. This new focus meant financial and military support for anticommunist governments around the world and a significantly increased commitment to military spending. In 1982, Reagan became the first U.S. president to address the British Parliament, where he laid out the heart of his foreign policy: "peace through strength." How do Reagan's vision of the Cold War and his articulation of the idea of freedom compare with the nuclear consensus of the 1950s? What has changed, and what has remained constant? What role does he see for nuclear energy in this renewed consensus? What vision of civil society does Reagan express?

We have not inherited an easy world. If developments like the Industrial Revolution, which began here in England, and the gifts of science and technology have made life much easier for us, they have also made it more dangerous. There are threats now to our freedom,

Excerpted from Ronald Reagan, "Address to Members of the British Parliament," June 8, 1982, Ronald Reagan Presidential Library, www.reagan.utexas.edu/archives/speeches/1982/60882a.htm.

indeed to our very existence, that other generations could never even have imagined.

There is first the threat of global war. No President, no Congress, no Prime Minister, no Parliament can spend a day entirely free of this threat. And I don't have to tell you that in today's world the existence of nuclear weapons could mean, if not the extinction of mankind, then surely the end of civilization as we know it. That's why negotiations on intermediate-range nuclear forces now underway in Europe and the START talks—Strategic Arms Reduction Talks—which will begin later this month, are not just critical to American or Western policy; they are critical to mankind. Our commitment to early success in these negotiations is firm and unshakable, and our purpose is clear: reducing the risk of war by reducing the means of waging war on both sides.

At the same time there is a threat posed to human freedom by the enormous power of the modern state. History teaches the dangers of government that overreaches—political control taking precedence over free economic growth, secret police, mindless bureaucracy, all combining to stifle individual excellence and personal freedom.

Now, I'm aware that among us here and throughout Europe there is legitimate disagreement over the extent to which the public sector should play a role in a nation's economy and life. But on one point all of us are united—our abhorrence of dictatorship in all its forms, but most particularly totalitarianism and the terrible inhumanities it has caused in our time—the great purge, Auschwitz and Dachau, the Gulag, and Cambodia. . . .

If history teaches anything it teaches self-delusion in the face of unpleasant facts is folly. We see around us today the marks of our terrible dilemma—predictions of doomsday, antinuclear demonstrations, an arms race in which the West must, for its own protection, be an unwilling participant. At the same time we see totalitarian forces in the world who seek subversion and conflict around the globe to further their barbarous assault on the human spirit. What, then, is our course? Must civilization perish in a hail of fiery atoms? Must freedom wither in a quiet, deadening accommodation with totalitarian evil? . . .

No, democracy is not a fragile flower. Still it needs cultivating. If the rest of this century is to witness the gradual growth of freedom and democratic ideals, we must take actions to assist the campaign for democracy.

Some argue that we should encourage democratic change in right-wing dictatorships, but not in Communist regimes. Well, to accept this preposterous notion—as some well-meaning people have—is to invite the argument that once countries achieve a nuclear capability, they should be allowed an undisturbed reign of terror over their own citizens. We reject this course.

As for the Soviet view, Chairman [Leonid] Brezhnev repeatedly has stressed that the competition of ideas and systems must continue and that this is entirely consistent with relaxation of tensions and peace....

We cannot ignore the fact that even without our encouragement there has been and will continue to be repeated explosions against repression and dictatorships. The Soviet Union itself is not immune to this reality. Any system is inherently unstable that has no peaceful means to legitimize its leaders. In such cases, the very repressiveness of the state ultimately drives people to resist it, if necessary, by force.

While we must be cautious about forcing the pace of change, we must not hesitate to declare our ultimate objectives and to take concrete actions to move toward them. We must be staunch in our conviction that freedom is not the sole prerogative of a lucky few, but the inalienable and universal right of all human beings. So states the United Nations Universal Declaration of Human Rights, which, among other things, guarantees free elections.

The objective I propose is quite simple to state: to foster the infrastructure of democracy, the system of a free press, unions, political parties, universities, which allows a people to choose their own way to develop their own culture, to reconcile their own differences through peaceful means.

This is not cultural imperialism, it is providing the means for genuine self-determination and protection for diversity. Democracy already flourishes in countries with very different cultures and historical experiences. It would be cultural condescension, or worse, to say that any people prefer dictatorship to democracy. Who would voluntarily choose not to have the right to vote, decide to purchase government propaganda handouts instead of independent newspapers, prefer government to worker-controlled unions, opt for land to be owned by the state instead of those who till it, want government repression of religious liberty, a single political party instead of a free choice, a rigid cultural orthodoxy instead of democratic tolerance and diversity? . . .

I have discussed on other occasions . . . the elements of Western policies toward the Soviet Union to safeguard our interests and protect the peace. What I am describing now is a plan and a hope for the long term—the march of freedom and democracy which will leave Marxism-Leninism on the ash-heap of history as it has left other tyrannies which stifle the freedom and muzzle the self-expression of the people. And that's why we must continue our efforts to strengthen NATO [the North Atlantic Treaty Alliance] even as we move forward . . . in the negotiations on intermediate-range forces and our proposal for a one-third reduction in strategic ballistic missile warheads.

Our military strength is a prerequisite to peace, but let it be clear we maintain this strength in the hope it will never be used, for the ultimate determinant in the struggle that's now going on in the world will not be bombs and rockets, but a test of wills and ideas, a trial of spiritual resolve, the values we hold, the beliefs we cherish, the ideals to which we are dedicated.

NUCLEAR WASTE POLICY ACT OF 1982

In the early 1980s the Nuclear Regulatory Commission and the nuclear energy industry still struggled with the question of what to do with spent fuel. Commercial reprocessing had ceased in the late 1970s, and spent fuel accumulated at reactors around the country. In an attempt to deal with this problem, Congress passed the Nuclear Waste Policy Act of 1982, laying out a process for the development of a geologic repository—an underground facility that would sequester high-level radioactive wastes for thousands of years. The bill proved contentious from the start, as it attempted to manage a host of difficult technical, ethical, and economic factors. Where did Congress suggest that waste should, and should not, go? Who would pay for its storage? What kind of provision does the act make for public comment and local control of the decision-making process?

. . . .

SEC. 111. (a) FINDINGS.—The Congress finds that—

(1) radioactive waste creates potential risks and requires safe and environmentally acceptable methods of disposal;

(2) a national problem has been created by the accumulation of (A) spent nuclear fuel from nuclear reactors; and (B) radioactive waste from (i) reprocessing of spent nuclear fuel; (ii) activities related to medical research, diagnosis, and treatment; and (iii) other sources;

Excerpted from the *Nuclear Waste Policy Act of 1982*, U.S. Code 42 (1982), 10101 et seq.

(3) Federal efforts during the past 30 years to devise a permanent solution to the problems of civilian radioactive waste disposal have not been adequate;

(4) while the Federal Government has the responsibility to provide for the permanent disposal of high-level radioactive waste and such spent nuclear fuel as may be disposed of in order to protect the public health and safety and the environment, the costs of such disposal should be the responsibility of the generators and owners of such waste and spent fuel;

(5) the generators and owners of high-level radioactive waste and spent nuclear fuel have the primary responsibility to provide for, and the responsibility to pay the costs of, the interim storage of such waste and spent fuel until such waste and spent fuel is accepted by the Secretary of Energy in accordance with the provisions of this Act;

(6) State and public participation in the planning and development of repositories is essential in order to promote public confidence in the safety of disposal of such waste and spent fuel; and

(7) high-level radioactive waste and spent nuclear fuel have become major subjects of public concern, and appropriate precautions must be taken to ensure that such waste and spent fuel do not adversely affect the public health and safety and the environment for this or future generations.

(b) PURPOSES.—The purposes of this subtitle are—

(1) to establish a schedule for the siting, construction, and operation of repositories that will provide a reasonable assurance that the public and the environment will be adequately protected from the hazards posed by high-level radioactive waste and such spent nuclear fuel as may be disposed of in a repository;

(2) to establish the Federal responsibility, and a definite Federal policy, for the disposal of such waste and spent fuel;

(3) to define the relationship between the Federal Government and the State governments with respect to the disposal of such waste and spent fuel; and

(4) to establish a Nuclear Waste Fund, composed of payments made by the generators and owners of such waste and spent fuel, that will ensure that the costs of carrying out activities relating to the disposal of such waste and spent fuel will be borne by the persons responsible for generating such waste and spent fuel.

RECOMMENDATION OF CANDIDATE SITES FOR SITE CHARACTERIZATION

SEC. 112. (a) GUIDELINES.—Not later than 180 days after the date of the enactment of this Act, the Secretary [of Energy], following consultation with the Council on Environmental Quality, the Administrator of the Environmental Protection Agency, the Director of the United States Geological Survey, and interested Governors, and the concurrence of the Commission shall issue general guidelines for the recommendation of sites for repositories. Such guidelines shall specify detailed geologic considerations that shall be primary criteria for the selection of sites in various geologic media. Such guidelines shall specify factors that qualify or disqualify any site from development as a repository, including factors pertaining to the location of valuable natural resources, hydrology, geophysics, seismic activity, and atomic energy defense activities, proximity to water supplies, proximity to populations, the effect upon the rights of users of water, and proximity to components of the National Park System, the National Wildlife Refuge System, the National Wild and Scenic Rivers System, the National Wilderness Preservation System, or National Forest Lands. Such guidelines shall take into consideration the proximity to sites where high-level radioactive waste and spent nuclear fuel is generated or temporarily stored and the transportation and safety factors involved in moving such waste to a repository. Such guidelines shall specify population factors that will disqualify any site from development as a repository if any surface facility of such repository would be located (1) in a highly populated area;

or (2) adjacent to an area 1 mile by 1 mile having a population of not less than 1,000 individuals. Such guidelines also shall require the Secretary to consider the cost and impact of transporting to the repository site the solidified high-level radioactive waste and spent fuel to be disposed of in the repository and the advantages of regional distribution in the siting of repositories. Such guidelines shall require the Secretary to consider the various geologic media in which sites for repositories may be located and, to the extent practicable, to recommend sites in different geologic media. The Secretary shall use guidelines established under this subsection in considering candidate sites for recommendation under subsection (b). The Secretary may revise such guidelines from time to time, consistent with the provisions of this subsection.

(b) RECOMMENDATION BY SECRETARY to the PRESIDENT.— (1)(A) Following the issuance of guidelines under subsection (a) and consultation with the Governors of affected States, the Secretary shall nominate at least 5 sites that he determines suitable for site characterization for selection of the first repository site.

(B) Subsequent to such nomination, the Secretary shall recommend to the President 3 of the nominated sites not later than January 1, 1985 for characterization as candidate sites. . . .

(D) Each nomination of a site under this subsection shall be accompanied by an environmental assessment, which shall include a detailed statement of the basis for such recommendation and of the probable impacts of the site characterization activities planned for such site, and a discussion of alternative activities relating to site characterization that may be undertaken to avoid such impacts. . . .

(F) Each environmental assessment prepared under this paragraph shall be made available to the public.

(G) Before nominating a site, the Secretary shall notify the Governor and legislature of the State in which such site is located, or the governing body of the affected Indian tribe where such site is located, as the case may be, of such nomination and the basis for such nomination.

(2) Before nominating any site the Secretary shall hold public hearings in the vicinity of such site to inform the residents of the area in which such site is located of the proposed nomination of such site and to receive their comments. At such hearings, the Secretary shall also solicit and receive any recommendations of such residents with respect to issues that should be addressed in the environmental assessment described in paragraph (1) and the site characterization plan described in section 113(b)(1). . . .

(c) PRESIDENTIAL REVIEW OF RECOMMENDED CANDIDATE SITES.—(1) The President shall review each candidate site recommendation made by the Secretary under subsection (b). Not later than 60 days after the submission by the Secretary of a recommendation of a candidate site, the President, in his discretion, may either approve or disapprove such candidate site, and shall transmit any such decision to the Secretary and to either the Governor and legislature of the State in which such candidate site is located, or the governing body of the affected Indian tribe where such candidate site is located, as the case may be. If, during such 60-day period, the President fails to approve or disapprove such candidate site, or fails to invoke his authority under paragraph (2) to delay his decision, such candidate site shall be considered to be approved, and the Secretary shall notify such Governor and legislature, or governing body of the affected Indian tribe, of the approval of such candidate site by reason of the inaction of the President.

JONATHAN SCHELL

THE FATE OF THE EARTH, 1982

The renewed Cold War tensions of the 1980s led to a corresponding uptick in concern about the arms race and fear of a nuclear war. Journalist and author Jonathan Schell played a key role in this growing concern when he published a series of articles on the arms race in the *New Yorker.* He subsequently gathered the articles into a best-selling and controversial book, *The Fate of the Earth.* The book helped to galvanize the peace and antinuclear movements. Schell focused on the potential impacts of a nuclear war from ecological and metaphysical perspectives. What does he see as the human place in nature? How does he invoke ecological rhetoric to make his point? On who does he place the obligation to act, and why? What does this excerpt suggest about such topics as death and extinction, science and progress, and human nature?

....

Four and a half billion years ago, the earth was formed. Perhaps a half billion years after that, life arose on the planet. For the next four billion years, life became steadily more complex, more varied, and more ingenious, until, around a million years ago, it produced mankind—the most complex and ingenious species of them all. Only six or seven thousand years ago—a period that is to the history of the earth as less than a minute is to a year—civilization emerged, enabling us to build up a human world, and to add to the marvels of evolution marvels of our own: marvels of art, of science, of social organization, of spiritual attainment. But, as we built higher and higher, the evolutionary foundation beneath our feet became more and more shaky, and now, in spite of all we have learned and achieved—or, rather, because of it—we hold this entire terrestrial creation hostage to nuclear destruction, threatening to hurl it back into the inanimate darkness from which it came. And this threat of self-destruction and planetary destruction is not something that we will pose one day in the future, if we fail to take certain precautions; it is here now, hanging over the heads of all of us at every moment. The machinery of destruction is complete, poised on a hair trigger, waiting for the "button" to be "pushed" by some misguided or deranged human being or for some faulty computer chip to send out the instruction to fire. That so much should be balanced on so fine a point—that the fruit of four and a half billion years can be undone in a careless moment—is a fact against which belief rebels. And there is another, even vaster measure of the loss, for stretching ahead from our present are more billions of years of life on earth, all of which can be filled not only with human life but with human civilization. The procession of generations that extends onward from our present leads far, far beyond the line of our sight, and, compared with these stretches of human time, which exceed the whole history of the earth up to now, our brief civilized moment is almost infinitesimal. And yet we threaten, in the name of our transient aims and fallible convictions, to foreclose it all. If our species does destroy itself, it will be a death in the cradle—a case of infant mortality. The disparity between the cause and the effect of our peril is so great that our minds seem all but powerless to encompass it. In addition, we are so fully enveloped by that which is menaced, and so deeply and passionately immersed in its events, which are the events of our lives, that we hardly know how to get far enough away from it to see it in its entirety. It is as though life itself were one huge distraction, diverting our attention from the peril to life. In its apparent durability, a world menaced with imminent doom is in a way deceptive. It is almost an illusion. Now we are sitting at the breakfast table drinking our coffee and reading the newspaper, but in a moment we may be inside a fireball whose temperature is tens of thousands of degrees. Now we are on our way to work, walking through the city streets, but in a moment we may be standing on an empty plain under a darkened sky looking for the charred remnants of our children. Now we are alive, but in a moment we may be dead. Now there is human life on earth, but in a moment it may be gone. . . .

In the face of this unprecedented global emergency, we have so far had no better idea than to heap up more and more warheads, apparently in the hope of so thoroughly paralyzing ourselves with terror that we will hold back from taking the final, absurd step. Considering the wealth of our achievement as a species, this response is unworthy of us. Only by a process of gradual debasement of our self-esteem can we have lowered our expectations to this point. For, of all the "modest hopes of human beings," the hope that mankind will survive is the most modest, since it only brings us to the threshold of all the other hopes. In entertaining it, we do not yet ask for justice, or for freedom, or for happiness, or for any of the other things that we may want in life. We do not even necessarily ask for our personal survival; we ask only that we *be survived*. We ask for assurance that when we die as individuals, as we know we must, mankind will live on. Yet once the peril of extinction is present, as it is for us now, the hope for human survival becomes the most tremendous hope, just because it is the foundation for all the other hopes, and in its absence every other hope will gradually wither and die. Life without the hope for human survival is a life of despair.

The death of our species resembles the death of an individual in its boundlessness, its blankness, its removal beyond experience, and its tendency to baffle human thought and feeling, yet as soon as one mentions the hope of survival the similarities are clearly at an end. For while individual death is inevitable, extinction can be avoided; while every person must die, mankind can be saved. Therefore, while reflection on death may lead to resignation and acceptance, reflection on extinction must lead to exactly the opposite response: to arousal, rejection, indignation, and action. Extinction is not something to contemplate, it is something to rebel against. To point this out might seem like stating the obvious if it were not that on the whole the world's reaction to the peril of extinction has been one of numbness and inertia, much as though extinction were as inescapable as death is. Even today, the official response to the sickening reality before us is conditioned by a grim fatalism, in which the hope of ridding the world of nuclear weapons, and thus of surviving as a species, is all but ruled out of consideration as "utopian" or "extreme"—as though it were "radical" merely to want to go on living and to want one's descendants to be born. And yet if one gives up these aspirations one has given up on everything. As a species, we have as yet done nothing to save ourselves. The slate of action is blank. We have organizations for the preservation of almost everything in life that we want but no organization for the preservation of mankind. People seem to have decided that our collective will is too weak or flawed to rise to this occasion. They see the violence that has saturated human history, and conclude that to practice violence is innate in our species. They find the perennial hope that peace can be brought to the earth once and for all a delusion of the well-meaning who have refused to face the "harsh realities" of international life—the realities of self-interest, fear, hatred, and aggression. They have concluded that these realities are eternal ones, and this conclusion defeats at the outset any hope of taking the actions necessary for survival. Looking at the historical record, they ask what has changed to give anyone confidence that humanity can break with its violent past and act with greater restraint. The answer, of course, is that everything has changed. To the old "harsh realities" of international life has been added the immeasurably harsher new reality of the peril of extinction. To the old truth that all men are brothers has been added the inescapable new truth that not only on the moral but also on the physical plane the nation that practices aggression will itself die. This is the law of the doctrine of nuclear deterrence—the doctrine of "mutual assured destruction"—which "assures" the destruction of the society of the attacker. And it is also the law of the natural world, which, in its own version of deterrence, supplements the oneness of mankind with a oneness of nature, and guarantees that when the attack rises above a certain level the attacker will be engulfed in the general ruin of the global ecosphere. To the obligation to honor life is now added the sanction that if we fail in our obligation life will actually be taken away from us, individually and collectively. Each of us will die, and as we die we will see the world around us dying. Such imponderables as the sum of human life, the integrity of the terrestrial creation, and the meaning of time, of history, and of the development of life on earth, which were once left to contemplation and spiritual understanding, are now at stake in the political realm and demand a political response from every person. As political actors, we must, like the contemplatives before us, delve to the bottom of the world, and, Atlas-like, we must take the world on our shoulders.

RONALD REAGAN

"ADDRESS TO THE NATION ON DEFENSE AND NATIONAL SECURITY," 1983

President Ronald Reagan responded to the revitalized peace movement by reiterating his commitment to "peace through strength" and continuing a buildup of American military forces. He paired these goals with a new initiative in 1983, the Strategic Defense Initiative (SDI), which would create a defensive shield by mounting lasers on space-based satellites with the capability of shooting down missiles before they could hit their intended targets. Opponents—who dubbed the plan "Star Wars" after the popular movie franchise—doubted the viability of the technology, worried about the implications of defensive technologies for deterrence, and claimed that the program would jeopardize arms control negotiations. In the speech excerpted below, Reagan introduced SDI to the American public. He also directly addressed the nuclear freeze movement, which called on the United States and the Soviet Union to halt the testing, production, and deployment of nuclear weapons. The nuclear freeze movement gathered broad public support in many venues around the United States, including in Congress. How does Reagan assess the nuclear freeze, and in what way is SDI a response to it? How does he address concerns about the potentially destabilizing nature of SDI?

Excerpted from Ronald Reagan, "Address to the Nation on Defense and National Security," March 23, 1983, Ronald Reagan Presidential Library, www.reagan.utexas.edu/archives/speeches/1983/32383d.htm.

Since the dawn of the atomic age, we've sought to reduce the risk of war by maintaining a strong deterrent and by seeking genuine arms control. "Deterrence" means simply this: making sure any adversary who thinks about attacking the United States, or our allies, or our vital interests, concludes that the risks to him outweigh any potential gains. Once he understands that, he won't attack. We maintain the peace through our strength; weakness only invites aggression.

This strategy of deterrence has not changed. It still works. But what it takes to maintain deterrence has changed. It took one kind of military force to deter an attack when we had far more nuclear weapons than any other power; it takes another kind now that the Soviets, for example, have enough accurate and powerful nuclear weapons to destroy virtually all of our missiles on the ground. Now, this is not to say that the Soviet Union is planning to make war on us. Nor do I believe a war is inevitable—quite the contrary. But what must be recognized is that our security is based on being prepared to meet all threats. . . .

I know that all of you want peace, and so do I. I know too that many of you seriously believe that a nuclear freeze would further the cause of peace. But a freeze now would make us less, not more, secure and would raise, not reduce, the risks of war. It would be largely unverifiable and would seriously undercut our negotiations on arms reduction. It would reward the Soviets for their massive military buildup while preventing us from modernizing our aging and increasingly vulnerable forces. With their present margin of superiority, why should they agree to arms reductions knowing that we were prohibited from catching up?

Believe me, it wasn't pleasant for someone who had come to Washington determined to reduce government spending, but we had to move forward with the task of repairing our defenses or we would lose our ability to deter conflict now and in the future. We had to demonstrate to any adversary that aggression could not succeed, and that the only real solution was substantial, equitable, and effectively verifiable arms reduction—the kind we're working for right now in Geneva. . . .

This approach to stability through offensive threat has worked. We and our allies have succeeded in preventing nuclear war for more than three decades. In recent months, however, my advisers, including in particular the Joint Chiefs of Staff, have underscored the necessity to break out of a future that relies solely on offensive retaliation for our security.

Over the course of these discussions, I've become more and more deeply convinced that the human spirit must be capable of rising above dealing with other nations and human beings by threatening their existence. Feeling this way, I believe we must thoroughly examine every opportunity for reducing tensions and for introducing greater stability into the strategic calculus on both sides.

One of the most important contributions we can make is, of course, to lower the level of all arms, and particularly nuclear arms. We're engaged right now in several negotiations with the Soviet Union to bring about a mutual reduction of weapons. I will report to you a week from tomorrow my thoughts on that score. But let me just say, I'm totally committed to this course.

If the Soviet Union will join with us in our effort to achieve major arms reduction, we will have succeeded in stabilizing the nuclear balance. Nevertheless, it will still be necessary to rely on the specter of retaliation, on mutual threat. And that's a sad commentary on the human condition. Wouldn't it be better to save lives than to avenge them? Are we not capable of demonstrating our peaceful intentions by applying all our abilities and our ingenuity to achieving a truly lasting stability? I think we are. Indeed, we must.

After careful consultation with my advisers, including the Joint Chiefs of Staff, I believe there is a way. Let me share with you a vision of the future which offers hope. It is that we embark on a program to counter the awesome Soviet missile threat with measures that are defensive. Let us turn to the very strengths in technology that spawned our great industrial base and that have given us the quality of life we enjoy today.

What if free people could live secure in the knowledge that their security did not rest upon the threat of instant U.S. retaliation to deter a Soviet attack, that we could intercept and destroy strategic ballistic missiles before they reached our own soil or that of our allies?

I know this is a formidable, technical task, one that may not be accomplished before the end of this century. Yet, current technology has attained a level of sophistication where it's reasonable for us to begin this effort. It will take years, probably decades of effort on many fronts. There will be failures and setbacks, just as there will be successes and breakthroughs. And as we proceed, we must remain constant in preserving the nuclear deterrent and maintaining a solid capability for flexible response. But isn't it worth every investment necessary to free the world from the threat of nuclear war? We know it is. . . .

America does possess—now—the technologies to attain very significant improvements in the effectiveness of our conventional, non-nuclear forces. Proceeding boldly with these new technologies, we can significantly reduce any incentive that the Soviet Union may have to threaten attack against the United States or its allies.

As we pursue our goal of defensive technologies, we recognize that our allies rely upon our strategic offensive power to deter attacks against them. Their vital interests and ours are inextricably linked. Their safety and ours are one. And no change in technology can or will alter that reality. We must and shall continue to honor our commitments.

I clearly recognize that defensive systems have limitations and raise certain problems and ambiguities. If paired with offensive systems, they can be viewed as fostering an aggressive policy, and no one wants that. But with these considerations firmly in mind, I call upon the scientific community in our country, those who gave us nuclear weapons, to turn their great talents now to the cause of mankind and world peace, to give us the means of rendering these nuclear weapons impotent and obsolete.

Tonight . . . I'm taking an important first step. I am directing a comprehensive and intensive effort to define a long-term research and development program to begin to achieve our ultimate goal of eliminating the threat posed by strategic nuclear missiles. This could pave the way for arms control measures to eliminate the weapons themselves. We seek neither military superiority nor political advantage. Our only purpose—one all people share—is to search for ways to reduce the danger of nuclear war.

CARL SAGAN

"THE NUCLEAR WINTER," 1983

Astronomer Carl Sagan was one of the most recognizable scientific voices of the late twentieth century, familiar to many Americans for writing and narrating the television series *Cosmos*—one of the most popular shows in the history of public television—and authoring many books and articles. Sagan took a public stance against the arms race in the early 1980s. In 1983 he and several colleagues authored the nuclear winter hypothesis—the theory that the dust and smoke generated by even a limited nuclear war might blot out the sun, radically depressing temperatures and endangering food supplies around the globe, ultimately leading to devastating famines. Sagan published the hypothesis in the popular magazine *Parade*—reprinted below—and also in the prominent academic journal *Science*. Sagan very explicitly tried to use his scientific credentials in a political way, responding to the Reagan administration's aggressive rhetoric and support for the Strategic Defense Initiative, both of which seemed to suggest the concept of a "winnable" nuclear war. Partly because of Sagan's activist stance, the nuclear winter hypothesis received a cool reception in the scientific community and generated a lively scholarly debate in subsequent decades. How does Sagan use science to support his argument? What does he want to achieve? How does his use of science compare to other documents in this book?

Except for fools and madmen, everyone knows that nuclear war would be an unprecedented human catastrophe. A more or less typical strategic warhead has a yield of 2 megatons, the explosive equivalent of 2 million tons of TNT. But 2 million tons of TNT is about the same as all the bombs exploded in World War II—a single bomb with the explosive power of the entire Second World War but compressed into a few seconds of time and an area 30 or 40 miles across. . . .

In a 2-megaton explosion over a fairly large city, buildings would be vaporized, people reduced to atoms and shadows, outlying structures blown down like matchsticks and raging fires ignited. And if the bomb were exploded on the ground, an enormous crater, like those that can be seen through a telescope on the surface of the Moon, would be all that remained where midtown once had been. There are now more than 50,000 nuclear weapons, more than 13,000 megatons of yield, deployed in the arsenals of the United States and the Soviet Union—enough to obliterate a million Hiroshimas.

But there are fewer than 3000 cities on the Earth with populations of 100,000 or more. You cannot find anything like a million Hiroshimas to obliterate. Prime military and industrial targets that are far from cities are comparatively rare. Thus, there are vastly more nuclear weapons than are needed for any plausible deterrence of a potential adversary.

Nobody knows, of course, how many megatons would be exploded in a real nuclear war. There are some who think that a nuclear war can be "contained," bottled up before it runs away to involve much of the world's arsenals. But a number of detailed analyses, war games run by the U.S. Department of Defense, and official Soviet pronouncements all indicate that this containment may be too much to hope for. Once the bombs begin exploding, communications failures, disorganization, fear, the necessity of making in minutes decisions affecting the fates of millions, and the immense psychological burden of knowing that your own loved ones may already have been destroyed are likely to result in a nuclear paroxysm. Many investigations, including a number of studies for the U.S. government, envision the explosion of 5000 to 10,000 megatons—the detonation of tens of thousands of nuclear weapons that now sit quietly, inconspicuously, in missile silos, submarines and long-range bombers, faithful servants awaiting orders.

The World Health Organization, in a recent detailed study chaired by Sune K. Bergstrom (the 1982 Nobel laureate in physiology and medicine), concludes that 1.1 billion people would be killed outright in such a nuclear war, mainly in the United States, the Soviet Union, Europe, China and Japan. An additional 1.1 billion people would suffer serious injuries and radiation sickness, for which medical help would be unavailable. It thus seems possible that more than 2 billion people—almost half of all the humans on Earth—would be destroyed in the immediate aftermath of a global thermonuclear war. This would represent by far the greatest disaster in the history of the human species and, with no other adverse effects, would probably be enough to reduce at least the Northern Hemisphere to a state of prolonged agony and barbarism. Unfortunately, the real situation would be much worse.

In technical studies of the consequences of nuclear weapons explosions, there has been a dangerous tendency to underestimate the results. This is partly due to a tradition of conservatism which generally works well in science but which is of more dubious applicability when the lives of billions of people are at stake. In the Bravo test of March 1, 1954, a 15-megaton thermonuclear bomb was exploded on Bikini Atoll. It had about double the yield expected, and there was an unanticipated last-minute shift in the wind direction. As a result, deadly radioactive fallout came down on Rongelap in the Marshall Islands, more than 200 kilometers away. Almost all the children on Rongelap subsequently developed thyroid nodules and lesions, and other long-term medical problems, due to the radioactive fallout.

Likewise, in 1973, it was discovered that high-yield airbursts will chemically burn the nitrogen in the upper air, converting it into oxides of nitrogen; these, in turn, combine with and destroy the protective ozone in the Earth's stratosphere. The surface of the Earth is shielded from deadly solar ultraviolet radiation by a layer of ozone so tenuous that, were it brought down to sea level, it would be only 3 millimeters thick. Partial destruction of this ozone layer can have serious consequences for the biology of the entire planet.

These discoveries, and others like them, were made by chance. They were largely unexpected. And now another consequence—by far the most dire—has been uncovered, again more or less by accident.

The U.S. Mariner 9 spacecraft, the first vehicle to orbit another planet, arrived at Mars in late 1971. The planet was enveloped in a global dust storm. As the fine particles slowly fell out, we were able to measure temperature changes in the atmosphere and on the surface. Soon it became clear what had happened:

The dust, lofted by high winds off the desert into the upper Martian atmosphere, had absorbed the incoming sunlight and prevented much of it from reaching the ground. Heated by the sunlight, the dust warmed the adjacent air. But the surface, enveloped in partial darkness, became much chillier than usual. Months later, after the dust fell out of the atmosphere, the upper air cooled and the surface warmed, both returning to their normal conditions. We were able to calculate accurately, from how much dust there was in the atmosphere, how cool the Martian surface ought to have been.

Afterwards, I and my colleagues, James B. Pollack and Brian Toon of NASA's Ames Research Center, were eager to apply these insights to Earth. In a volcanic explosion, dust aerosols are lofted into the high atmosphere. We calculated by how much the Earth's global temperature should decline after a major volcanic explosion and found that our results (generally a fraction of a degree) were in good accord with actual measurements. Joining forces with Richard Turco, who has studied the effects of nuclear weapons for many years, we then began to turn our attention to the climatic implications of nuclear war. [The scientific paper, "Global Atmospheric Consequences of Nuclear War," is written by R. P. Turco, O. B. Toon, T. P. Ackerman, J. B. Pollack and Carl Sagan. From the last names of the authors, this work is generally referred to as "TTAPS."]

We knew that nuclear explosions, particularly groundbursts, would lift an enormous quantity of fine soil particles into the atmosphere (more than 100,000 tons of fine dust for every megaton exploded in a surface burst). Our work was further spurred by Paul Crutzen of the Max Planck Institute for Chemistry in Mainz, West Germany, and by John Birks of the University of Colorado, who pointed out that huge quantities of smoke would be generated in the burning of cities and forests following a nuclear war.

Groundbursts—at hardened missile silos, for example—generate fine dust. Airbursts—over cities and unhardened military installations—make fires and therefore smoke. The amount of dust and soot generated depends on the conduct of the war, the yields of the weapons employed and the ratio of groundbursts to airbursts. So we ran computer models for several dozen different nuclear war scenarios. Our baseline case, as in many other studies, was a 5000-megaton war with only a modest fraction of the yield (20 percent) expended on urban or industrial targets. Our job, for each case, was to follow the dust and smoke generated, see how much sunlight was absorbed and by how much the temperatures changed, figure out how the particles spread in longitude and latitude, and calculate how long before it all fell out of the air back onto the surface. Since the radioactivity would be attached to these same fine particles, our calculations also revealed the extent and timing of the subsequent radioactive fallout.

Some of what I am about to describe is horrifying. I know, because it horrifies me. There is a tendency—psychiatrists call it "denial"—to put it out of our minds, not to think about it. But if we are to deal intelligently, wisely, with the nuclear arms race, then we must steel ourselves to contemplate the horrors of nuclear war.

The results of our calculations astonished us. In the baseline case, the amount of sunlight at the ground was reduced to a few percent of normal—much darker, in daylight, than in a heavy overcast and too dark for plants to make a living from photosynthesis. At least in the Northern Hemisphere, where the great preponderance of strategic targets lies, a deadly gloom would persist for months.

Even more unexpected were the temperatures calculated. In the baseline case, land temperatures, except for narrow strips of coastline, dropped to minus 25° Celsius (minus 13° Fahrenheit) and stayed below freezing for months—even for a summer war. (Because the atmospheric structure becomes much more stable as the upper atmosphere is heated and the lower air is cooled, we may have severely underestimated how long the cold and the dark would last.) The oceans, a significant heat reservoir, would not freeze, however, and a major ice age would probably not be triggered. But because the temperatures would drop so catastrophically, virtually all crops and farm animals, at least in the Northern Hemisphere, would be destroyed, as would most varieties of uncultivated or undomesticated food supplies. Most of the human survivors would starve.

In addition, the amount of radioactive fallout is much more than expected. Many previous calculations simply ignored the intermediate time-scale fallout. That is, calculations were made for the prompt fallout—the plumes of radioactive debris blown downwind from each target—and for the long-term fallout, the fine radioactive particles lofted into the stratosphere that would descend about a year later, after most of the radioactivity had decayed. However, the radioactivity carried into the upper atmosphere (but not as high as the stratosphere) seems to have been largely forgotten. We found for the baseline case that roughly 30 percent of the land at northern midlatitudes could receive a radioactive dose greater than 250 rads, and that about 50 percent of northern midlatitudes could receive a dose greater than 100 rads. A 100-rad dose is the equivalent of about 1000 medical X-rays. A 400-rad dose will, more likely than not, kill you.

The cold, the dark and the intense radioactivity, together lasting for months, represent a severe assault on our civilization and our species. Civil and sanitary services would be wiped out. Medical facilities, drugs, the most rudimentary means for relieving the vast human suffering, would be unavailable. Any but the most elaborate shelters would be useless, quite apart from the question of what good it might be to emerge a few months later. Synthetics burned in the destruction of the cities would produce a wide variety of toxic gases, including carbon monoxide, cyanides, dioxins and furans. After the dust and soot settled out, the solar ultraviolet flux would be much larger than its present value.

Immunity to disease would decline. Epidemics and pandemics would be rampant, especially after the billion or so unburied bodies began to thaw. Moreover, the combined influence of these severe and simultaneous stresses on life are likely to produce even more adverse consequences—biologists call them synergisms—that we are not yet wise enough to foresee.

So far, we have talked only of the Northern Hemisphere. But it now seems—unlike the case of single nuclear weapons test—that in a real nuclear war, the heating of the vast quantities of atmospheric dust and soot in northern midlatitudes will transport these fine particles toward and across the Equator. We see just this happening in Martian dust storms. The Southern Hemisphere would experience effects that, while less severe than in the Northern Hemisphere, are nevertheless extremely ominous. The illusion with which some people in the Northern Hemisphere reassure themselves—catching an Air New Zealand flight in a time of serious international crisis, or the like—is now much less tenable, even on the narrow issue of personal survival for those with the price of a ticket.

But what if nuclear wars can be contained, and much less than 5000 megatons is detonated? Perhaps the greatest surprise in our work was that even small nuclear wars can have devastating climatic effects. We considered a war in which a mere 100 megatons were exploded, less than one percent of the world arsenals, and only in low-yield airbursts over cities. This scenario, we found, would ignite thousands of fires, and the smoke from these fires alone would be enough to generate an epoch of cold and dark almost as severe as in the 5000-megaton case. The threshold for what Richard Turco has called The Nuclear Winter is very low.

Could we have overlooked some important effect? The carrying of dust and soot from the Northern to the Southern Hemisphere (as well as more local atmospheric circulation) will certainly thin the clouds out over the Northern Hemisphere. But, in many cases, this thinning would be insufficient to render the climatic consequences tolerable—and every time it got better in the Northern Hemisphere, it would get worse in the Southern.

Our results have been carefully scrutinized by more than 100 scientists in the United States, Europe and the Soviet Union. There are still arguments on points of detail. But the overall conclusion seems to be agreed upon: There are severe and previously unanticipated global consequences of nuclear war—subfreezing temperatures in a twilit radioactive gloom lasting for months or longer.

Scientists initially underestimated the effects of fallout, were amazed that nuclear explosions in space disabled distant satellites, had no idea that the fireballs from high-yield thermonuclear explosions could deplete the ozone layer and missed altogether the possible climatic effects of nuclear dust and smoke. What else have we overlooked?

Nuclear war is a problem that can be treated only theoretically. It is not amenable to experimentation. Conceivably, we have left something important out of our analysis, and the effects are more modest than we calculate. On the other hand, it is also possible—and, from previous experience, even likely—that there are further adverse effects that no one has yet been wise enough to recognize. With billions of lives at stake, where does conservatism lie—in assuming that the results will be better than we calculate, or worse?

Many biologists, considering the nuclear winter that these calculations describe, believe they carry somber implications for life on Earth. Many species of plants and animals would become extinct. Vast numbers of surviving humans would starve to death. The delicate ecological relations that bind together organisms on Earth in a fabric of mutual dependency would be torn, perhaps irreparably. There is little question that our global civilization would be destroyed. The human population would be reduced to prehistoric levels, or less. Life for any survivors would be extremely hard. And there seems to be a real possibility of the extinction of the human species.

It is now almost 40 years since the invention of nuclear weapons. We have not yet experienced a global thermonuclear war—although on more than one occasion we have come tremulously close. I do not think our luck can hold forever. Men and machines are fallible, as recent events remind us. Fools and madmen do exist, and sometimes rise to power. Concentrating always on the near future, we have ignored the long-term consequences of our actions. We have placed our civilization and our species in jeopardy.

Fortunately, it is not yet too late. We can safeguard the planetary civilization and the human family if we so choose. There is no more important or more urgent issue.

OFFICE OF TECHNOLOGY ASSESSMENT

NUCLEAR POWER IN AN AGE OF UNCERTAINTY, 1984

The commercial nuclear power industry faced uncertain prospects in the mid-1980s. A distrustful public, a daunting regulatory environment, and lower-than-expected electricity growth in the 1970s caused utility companies to shy away from nuclear power. In the early 1980s two congressional committees asked the Office of Technology Assessment (OTA) to evaluate "how nuclear technology could evolve if the option is to be made more attractive to all the parties of concern." The OTA assessed the industry and suggested actions that the federal government could take to revive it. What economic, environmental, and political factors does the OTA consider? What assumptions about safety and environmental impact does the report make? How does the OTA explain the bleak outlook of the nuclear power industry, and who needs to act to improve the situation? Why did it consider public opinion to be so important, and how had it changed?

• •

OVERVIEW AND FINDINGS

Without significant changes in the technology, management, and level of public acceptance, nuclear power in the United States is unlikely to be expanded in this century beyond the reactors already under construction. Currently,

---

Excerpted from the Office of Technology Assessment, *Nuclear Power in an Age of Uncertainty* (Washington, D.C.: U.S. Congress, Office of Technology Assessment, 1984), xi–xii, 3–4, 211.

nuclear powerplants present too many financial risks as a result of uncertainties in electric demand growth, very high capital costs, operating problems, increasing regulatory requirements, and growing public opposition.

If all these risks were *inherent* to nuclear power, there would be little concern over its demise. *However, enough utilities have built nuclear reactors within acceptable cost limits, and operated them safely and reliably to demonstrate that the difficulties with this technology are not insurmountable.* Furthermore, there are national policy reasons why it could be highly desirable to have a nuclear option in the future if present problems can be overcome. Demand for electricity could grow to a level that would mandate the construction of many new powerplants. Uncertainties over the long-term environmental acceptability of coal and the adequacy of economical alternative energy sources are also great and underscore the potential importance of nuclear power.

Some of the problems that have plagued the present generation of reactors are due to the immaturity of the technology, and an underestimation by some utilities and their contractors of the difficulty of managing it. A major commitment was made to build large reactors before any had been completed. Many of these problems should not reoccur if new reactors are ordered. The changes that have been applied retroactively to existing reactors at great cost would be incorporated easily in new designs. Safety and reliability should be better. It is also likely that only those utilities that have adequately managed their nuclear projects would consider a new plant.

While important and essential, these improvements by themselves are probably not adequate to break the present impasse. Problems such as large cost overruns and subsequent rate increases, inadequate quality control, uneven reliability, operating mishaps, and accidents, have been numerous enough that the confidence of the public, investors, rate and safety regulators, and the utilities themselves is too low to be restored easily. *Unless this trust is restored, nuclear power will not be a credible energy option for this country. . . .*

Improvements in areas outside the technology itself must start with the management of existing reactors. *The Nuclear Regulatory Commission, as well as the [nuclear power industry], must ensure a commitment to excellence in construction and operation at the highest levels of nuclear utility management*. Improved training programs, tightened procedures, and heightened awareness of opportunities for improved safety and reliability would follow. If some utilities still prove unable to improve sufficiently, consideration could be given to the suspension of operating licenses until their nuclear operations reflect the required competence, perhaps by employing other utilities or service companies. Similarly, certification of utilities or operating companies could be considered as a prerequisite for permits for new plants in order to guarantee that only qualified companies would have responsibility. These are drastic steps, but they may be warranted because all nuclear reactors are hostage, in a sense, to the poorest performing units. *Public acceptance, which is necessary if the nuclear option is to revive, depends in part on* all *reactors performing reliably and safely*. . . .

The improvements in technology and operations described above should produce gains in public acceptance. Additional steps may be required, however, considering the current very low levels of support for more reactors. Addressing the concerns of the critics and providing assurance of a controlled rate of nuclear expansion could eliminate much of the reason for public disaffection. An important contribution to restoring public confidence could be made by a greater degree of openness by all parties concerned about the problems and benefits of nuclear power.

If progress can be made in all these areas, nuclear power would be much more likely to be considered when new electric-generation capacity is needed. Such progress will be difficult, however, because many divergent groups will have to work together and substantial technical and institutional change may be necessary. . . .

THE POLICY PROBLEM

The nuclear power industry is facing a period of extreme uncertainty. No nuclear plant now operating or still under active construction has been ordered since 1974, and every year since then has seen a decrease in the total utility commitment to nuclear power. By the end of this decade, almost all the projects still under construction will have been completed or canceled. Prospects for new domestic orders during the next few years are dim. . . .

If nuclear power were irrelevant to future energy needs, it would not be of great interest to policymakers. However, several other factors must be taken into account. While electric growth has been very low over the last decade (in fact, it was negative in 1982), there is no assurance that this trend will continue. Even growth that is quite modest by historical standards would mandate new plants—that have not been ordered yet—coming online in the 1990's. Replacement of aging plants will call for still more new generating capacity. The industrial capability already exists to meet new demand with nuclear reactors even if high electric growth resumes. In addition, reactors use an abundant resource. Oil is not a realistic option for new electric-generating plants because of already high costs and vulnerability to import disruptions which are likely to increase by the end of the century. Natural gas may also be too costly or unavailable for generating large quantities of electricity.

The use of coal can and will be expanded considerably. All the plausible growth projections considered in this study could be met entirely by coal. Such a dependence, however, would leave the Nation's electric system vulnerable to price increases and disruptions of supply. Furthermore, coal carries significant liabilities. The continued combustion of fossil fuels, especially coal, has the potential to release enough carbon dioxide to cause serious climatic changes. We do not know enough about this problem yet to say when it could happen or how severe it might be, but the possibility exists that even in the early 21st century it may become essential to reduce sharply the use of fossil fuels especially coal. Another potentially serious problem with coal is pollution in the form of acid rain, which already is causing considerable concern. Even with the strictest current control technology, a coal plant emits large quantities of the oxides of sulfur and nitrogen that are believed to be the primary source of the problem. There are great uncertainties in our understanding of this problem also, but the potential exists for large-scale coal combustion to become unacceptable or much more expensive due to tighter restrictions on emissions.

There are other possible alternatives to coal, of course. Improving the performance of existing powerplants would make more electricity available without building new capacity. Cogeneration and improved efficiency in the use of electricity also are equivalent to adding new supply. These approaches are likely to be the biggest contributors to

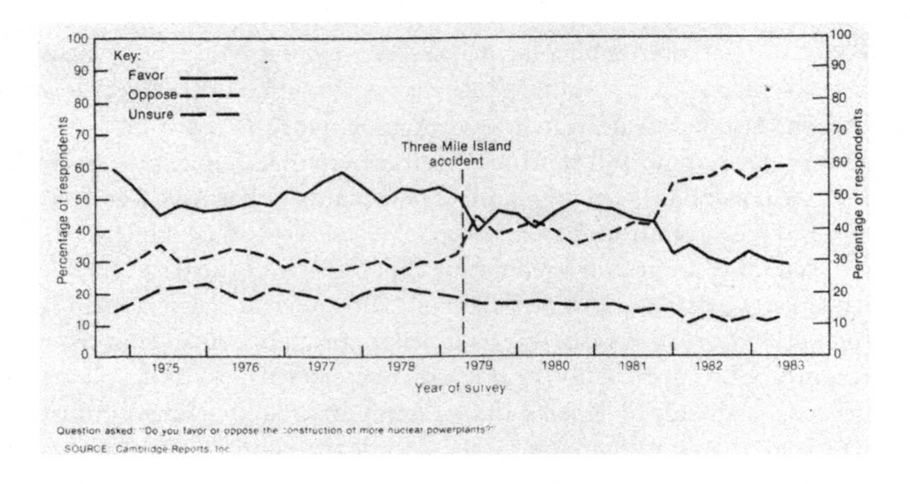

meeting new electric service requirements over the next few decades. Various forms of solar and geothermal energy also appear promising. Uncertainties of economics and applicability of these technologies, however, are too great to demonstrate that they will obviate the need for nuclear power over the next several decades.

Therefore, there may be good national-policy reasons for wanting to see the nuclear option preserved. However, the purpose of the preceding discussion is not to show that nuclear power necessarily is vital to this Nation's well-being. It is, rather, to suggest that there are conditions under which nuclear power would be the preferred choice, and that these conditions might not be recognized before the industry has lost its ability to supply reactors efficiently and expeditiously. If the nuclear option is foreclosed, it should at least happen with foresight, not by accident or neglect.

CAMPAIGN FOR A NUCLEAR FREE FUTURE

DISARMAMENT BEGINS AT HOME, CA. 1984

The antinuclear movement of the 1980s took many forms. While the nuclear freeze movement focused on arms control, a related organization called the Campaign for a Nuclear Free Future worked at the community level to encourage municipalities to declare themselves "nuclear free." This movement proved successful, with towns and cities ranging in size from Chicago and New York to Sykesville, Maryland, and Homer, Alaska, adopting some form of the resolution. What did advocates for this campaign call for? What motivated them to action? How does this document intersect with other statements about the way that nuclear energy shaped the contours of civil society?

DISARMAMENT BEGINS AT HOME

There's a fresh idea growing in the United States disarmament movement. People all across the country are organizing their communities to become Nuclear Free Zones.

Frustrated by the snails pace of international arms control negotiations, and alarmed by the world's headlong rush toward nuclear conflict, people all over the world have declared their opposition to the continued nuclear buildup. Millions of Americans have joined the Nuclear Freeze movement as a way of pressuring governments to end the arms race.

Excerpted from Campaign for a Nuclear Free Future pamphlet, *Disarmament Begins at Home,* ca. 1984, Marjorie Hannon Pifer Papers, Box 2, Folder "Other Anti-Nuclear Weapons Groups—Pamphlets," Wisconsin Historical Society Archives, Madison, Wisc. Courtesy of the Wisconsin Historical Society.

Now, without waiting for politicians and generals to act on our behalf, we have an opportunity to play a direct role in reversing the arms race: we can end our participation by making our communities Nuclear Free Zones.

A Nuclear Free Zone is an area or community that prohibits the research, testing, production, transportation, storage, and deployment of nuclear weapons. In addition, those systems and processes which support the nuclear arms race are dismantled and abolished, and community alternatives investigated. These might include nuclear power plants, nuclear waste dumps, and uranium mines. The local capital, skills, and resources made available as a result can be used to provide for housing, energy, health care, and other community needs

WHY A NUCLEAR FREE ZONE?

While national and international problems like the arms race and unsafe energy often seem too great and too far removed for us to have any effect, the roots of these problems lie close at hand, in our own communities. By establishing Nuclear Free Zones we are taking a direct and positive step toward reversing the arms race at the local level. Our action is a clear statement of nonviolent resistance, of non-cooperation with the forces of war and environmental destruction. In the process of becoming a Nuclear Free Zone, we can close down a uranium mine, halt nuclear research at a local university, or convert a local weapons facility to produce something useful for the community.

We can protect those areas which are presently nuclear free from falling victim to the cancer of militarism, and set those areas already infected on the road to recovery. By organizing Nuclear Free Zones, we can send a powerful message to world leaders that we will no longer submit to nuclear tyranny and that our communities should have a voice in determining how local resources should be used.

At the same time, we can go beyond what we don't want—nuclear facilities and arms production—to what we do want—communities that provide for our social needs, economic survival, and a decent quality of life for everyone. As we press local institutions to stop using local resources in support of the arms race, we can demand that these resources be used instead to meet local needs. A demand to end local participation in the nuclear industry will raise the issue of (and will struggle for) control over the decisions that affect a community's well-being.

COMMUNITY LINKS TO THE ARMS RACE

Every community in the U.S. is connected, directly or indirectly, to the nuclear arms race. The most obvious links are the factories that produce nuclear weapons and the missiles, planes, and ships that deliver them. But much more is involved. The nuclear fuel cycle begins with uranium mining and milling, mostly in the Southwest. Fuel is then enriched at plants in Tennessee, Ohio, and Kentucky. Nuclear weapons are tested, transported, and stored in communities around the country. Research centers on and off college campuses design and develop new weapons and delivery systems. Nuclear power plants, the commercial side of the nuclear industry, produce plutonium that can be used in developing weapons. Radioactive wastes from both nuclear weapons and energy programs are dumped at various sites around the country. And many of our local institutions hold financial investments in nuclear industries.

In addition, nearly every community in the U.S. is included in the federal government's Civil Defense plans. The elaborate (and unworkable) plans for massive evacuations of nuclear "risk areas" are part of a nuclear warfighting strategy, and foster the dangerous belief that a nuclear war can be won.

In a broader sense, we are connected to the nuclear arms race by the payment of federal war taxes and by cooperating in countless programs, processes and habits that promote militarism. All too many Congressional representatives help formulate policies which make nuclear war more likely, and local corporations provide the means for their implementation.

Every community feels the effects of the arms race, as funds and resources are diverted from social needs to the military. While record levels of military spending cripple the national economy, services and social programs are cut back or eliminated, and the quality of life in our communities deteriorates. All of our communities have felt the effects of rising unemployment and program cuts.

As more and more people educate themselves to the realities of the nuclear arms race, they have come to the conclusion that the problem does not rest only in Washington or Moscow, but in our own communities as well. Believing that disarmament begins at home, people have begun the process of disengaging their communities from the arms race. They are establishing Nuclear Free Zones.

The precise form of a Nuclear Free Zone depends on the nature of the community in which it is established. Some areas may focus on converting weapons facilities to socially useful, peacetime industries or press a local bank which invests in nuclear production to invest instead in the construction of a day care center. Others may seek to formally reject the federal government's Crisis Relocation Plan for Civil Defense. Some cities may pass zoning ordinances prohibiting certain types of economic and research activity associated with nuclear technology. Others may work to ban the transportation of radioactive wastes or other nuclear materials through their community. Congressional districts may instruct their representative to oppose funding for nuclear weapons. Still others may work to shut down a nuclear power plant, prevent new ones from being constructed and demand that utilities invest in conservation and safe energy.

In each case, community organizations can confront the nuclear offenders locally—whether they be corporations, banks, government agencies, or research facilities—and develop alternatives to nuclear weapons production, planning, and support. These alternatives would incorporate specific plans (and demands) for how the community's resources could be used to meet community needs rather than supporting nuclear proliferation.

BERNARD LOWN

"A PRESCRIPTION FOR HOPE," 1985

In 1980 the professional collaboration of two cardiologists—Russian Yevgeny Chazov and American Bernard Lown—led to the creation of the International Physicians for the Prevention of Nuclear War (IPPNW). The doctors vowed to fight against war as a result of their professional obligations to human health, labeling nuclear war the "final epidemic," and tried to foster conversation across the ideological divide of the Cold War. IPPNW grew quickly, totaling 145,000 members in forty nations by 1985. That year the IPPNW received the Nobel Peace Prize for its work. In this Nobel Peace Prize address, how does Lown connect environmental health to his responsibilities as a physician? How does he redefine the concept of "security," and how is this idea linked to his ideas about progress? In what ways does Lown challenge scientific perspectives, and in what ways does he reinforce them?

The advent of the nuclear age posed an unprecedented question: not whether war would exact yet more lives but whether war would preclude human existence altogether.

Every historic period has had its Cassandras. Our era is the first in which prophecies of doom stem from objective scientific analyses. Nearly a quarter of a century ago, a study by American physicians concluded that

Excerpted from Bernard Lown, "A Prescription for Hope," December 11, 1985, Nobel Prize Lecture, www.nobelprize.org/nobel_prizes/peace/laureates/1985/physicians-lecture.html.

medicine, which in past wars mitigated misery and saved lives, had nothing to offer following nuclear war. This conclusion was extrapolated from the destruction wrought by blast, fire and radiation on Hiroshima and Nagasaki. Astonishingly, nearly 40 years elapsed before scientists first discovered additional ecologic consequences. Nuclear war, they found, could blanket the sky with smoke, dust, and soot, creating a pall of all-pervasive darkness and frigid cold. The impact on climate could last for several years, not sparing the Southern Hemisphere.

But there is more. Since cities are enormous storehouses of combustible synthetics, raging fire storms would release into the air a Pandora's box of deadly toxins. When dust, poisons, and soot finally cleared, another plague would be visited on the unfortunate survivors; high levels of ultraviolet light caused by depletion of atmospheric ozone would take an additional toll. . . .

The nuclear threat haunts our age. Among the first to alert humanity to the peril were the physicists who let the atomic genie out of the bottle. Interestingly, though, the public is beginning to listen not to the military experts but to the physicians who are the custodians of public health. Now it may be argued that nuclear war is a social and political issue and we may address it only as concerned citizens. But we physicians have taken a sacred and ancient oath to assuage human misery and preserve life. This commitment imposes social and moral obligations for us to band together, to make our collective voices heard.

Furthermore, the medical profession cannot remain quiet in the face of the increasing diversion of scarce resources to the military compared to the meager efforts devoted to combating global poverty, malnutrition and disease. In 1984 world military spending exceeded 800,000 million dollars, or 100 million dollars every hour. This occurred at a time when life expectancy at birth in Africa is 30 years less than in Europe, when more than 40,000 children die daily from malnutrition and infection, when annually more than 3.5 million children die and an equal number are permanently crippled because they are denied inexpensive immunization. Two billion people have no access to a dependable and sanitary water supply. The litany of grief is long and painful to recite. Yet a single day's diversion of profligate military spending would diminish and even resolve many of these miseries. We are already living in the rubble of World War III. . . .

Perhaps the signal accomplishment of the IPPNW has been the broad-based, free-flowing dialogue between physicians of the two contending power blocs. We heed Einstein's words, "Peace cannot be kept by force. It can only be achieved by understanding." In a world riven with confrontation and strife, IPPNW has become a model for cooperation among physicians from East and West, from North and South. Paranoid fantasies of a dehumanized adversary cannot withstand the common pursuit of healing and preventing illness. Our success in forging such cooperation derives largely from an insistent avoidance of linkage with problems that have embittered relations between the great powers. We have resisted being sidetracked to other issues, no matter how morally lofty. Combating the nuclear threat has been our exclusive preoccupation, since we are dedicated to the proposition that to insure the conditions of life, we must prevent the conditions of death. Ultimately, we believe people must come to terms with the fact that the struggle is not between different national destinies, between opposing ideologies, but rather between catastrophe and survival. All nations share a linked destiny; nuclear weapons are their shared enemy.

The physicians' movement is contributing to a positive world outlook, rejecting the view that human life is merely the molecular unwinding of a dismal biologic clock. For the physician, whose role is to affirm life, optimism is a medical imperative. Even when the outcome is doubtful, a patient's hopeful attitude promotes well-being and frequently leads to recovery. Pessimism degrades the quality of life and jeopardizes the tomorrows yet to come. An affirmative world view is essential if we are to shape a more promising future.

The American poet Langston Hughes urged:

> "Hold on to dreams
> For if dreams die,
> Life is a broken winged bird
> [That] cannot fly."

We must hold fast to the dream that reason will prevail. The world today is full of anguish and dread. As great as is the danger, still greater is the opportunity. If science and technology have catapulted us to the brink of extinction, the same ingenuity has brought humankind to the boundary of an age of abundance.

Never before was it possible to feed all the hungry. Never before was it possible to shelter all the homeless. Never before was it possible to teach all the illiterates. Never before were we able to heal so many afflictions. For the first time science and medicine can diminish drudgery and pain.

ELIZABETH MACIAS

HIGH-LEVEL NUCLEAR WASTE ISSUES, 1987

By the mid-1980s the federal government appeared to make some headway in determining what to do with radioactive waste. The Department of Energy initiated construction of the Waste Isolation Pilot Plant in New Mexico to accept some of the waste generated from the production of nuclear weapons. In 1987, Congress revised the Nuclear Waste Policy Act of 1982 and selected Yucca Mountain in Nevada as the location for a geologic repository to receive commercial nuclear waste. The decision was controversial at the time—it was stridently opposed by the state's politicians—and has remained so since. Although the 1982 act mandated that the federal government would accept high-level wastes by 1998, no repository for commercial waste has as yet been constructed.

In the document excerpted below, Las Vegas city official Elizabeth Macias testifies before a Senate Committee, laying out objections to the transportation of radioactive wastes through her city and to their storage in Nevada. She raised concerns about transportation to both Yucca Mountain and to the Waste Isolation Pilot Plant, which stores military wastes and became the world's third operating geological repository when it finally opened in 1999. What factors does she think need to be considered, and why? How does she regard the process used to make decisions about waste storage and transportation?

Excerpted from Elizabeth Macias, *High-Level Nuclear Waste Issues*, Senate Committee on Environment and Public Works, Subcommittee on Nuclear Regulation, 100th Congress, 1st session, April 23, June 2–3, 18, 1987, 442–44.

Mr. Chairman, members of the subcommittee, I would like to thank you for this opportunity to express the City of Las Vegas' concerns regarding the transport of high-level nuclear waste. We commend the subcommittee for reviewing nuclear waste transportation at a time when citizens are demanding public policy that ensures the safest possible transport of these dangerous materials.

Las Vegas, the largest city in Nevada, serves as the center of the southern Nevada transportation network. Naturally, the transportation of hazardous materials is a major concern of our city. It is well known that elected city officials and administrators have been and will continue to be opposed to the transportation, transfer or storage of high-level nuclear waste and other hazardous materials in and throughout the City of Las Vegas.

This issue has been of growing concern due to the scheduled commencement in 1990 of shipments of Department of Energy–generated waste from the Nevada Test Site to the Waste Isolation Pilot Plant in Carlsbad, New Mexico. The plan is to truck six to ten shipments per year to the New Mexico facility. The route which has been tentatively selected uses Highway 95, through the heart of downtown Las Vegas, towards Boulder City and continues south on 95 to Needles, California.

While six to ten shipments per year does not seem considerable, the city is concerned that once these highways become targeted for such transport, a precedent will be established. Then, if Yucca Mountain does become the Nation's first nuclear waste repository, the tendency will be to send hundreds of shipments along the same route.

DOE's position on the transportation of nuclear waste states that "the overall goal is to reduce risk by reducing the amount of time the radioactive material is in transit." With the vast majority of nuclear waste being produced in the eastern United States, how does DOE justify shipping nuclear waste thousands of miles westward, if in fact their goal is to reduce travel time? Apparently, the Department of Energy gives little credence to transportation being a significant factor in site selection.

The only criteria that the DOE used to designate Waste Isolation Pilot Plant shipment routes were four-lane highways and shortest route. Using these factors, the Department of Energy originally scheduled their shipments to cross Hoover Dam, until they were finally prevailed upon to reconsider. The city is concerned that no analysis of the following factors has been considered: population characteristics; accident rates; high traffic and accident conditions; road conditions, such as pavement, shoulders and bridges; areas of extreme sensitivity, such as schools and hospitals; rail conditions, such as terrain and proximity to population centers; emergency response capabilities of communities bordering transportation corridors, and economic impact upon communities bordering shipping corridors.

In addition to the safety of our citizens, another major concern present is the impact of such shipments to our basic economy, tourism. Whether or not Yucca Mountain is selected as a site for the Nation's western repository, any transport of high-level nuclear waste through the city could adversely impact a person or corporation's decision to visit or site a facility in Las Vegas. In this case, the perception of danger is sufficient to damage both the city's tourism business and thwart economic diversification efforts.

The fear of a potential nuclear accident occurring along the proposed waste shipping routes translates into lost economic opportunities for the city and results in significant social impact as well for the community. The out-migration of residents, reduced property values and negative impact on the tourist industry affect not only the city's economy, but the entire State as well. The tax revenues generated by the tourism industry in Las Vegas provide a significant portion of the revenues collected by the entire State.

In addition, economic diversification efforts have included promoting Las Vegas as an ideal distribution center. Excellent highway and rail access to the western markets, coupled with the lack of a State inventory tax, have made Las Vegas a regional distribution center for several retail and wholesale operators. This growing business potential would be seriously impaired with the mere perception of danger associated with nuclear waste shipments.

The city feels that economic factors such as those I have mentioned must also be addressed and studied in detail before routes may be designated. . . .

The city feels that all State and local governments should have the right and be provided with the means to conduct their own independent studies for determining the safest routes through their communities. These independent findings must reflect a positive outcome rather than portray a situation that at best can be typified as potentially adverse. Funding for these studies should be provided in full at the Federal level. The selection of a designated route would be a collective effort, and the State and local governments should be permitted to provide input for every decision that affects them.

RONALD REAGAN

"ADDRESS TO THE 42ND SESSION OF THE UNITED NATIONS," 1987

In December 1987 the United States and the U.S.S.R. concluded a stunning treaty: the Intermediate-Range Nuclear Forces Treaty (INF). This treaty was the first arms control agreement that actually reduced the number of nuclear weapons—ballistic and cruise missiles with ranges between 300 to 3,400 miles. The INF treaty marked a sea change in the relationships between the two superpowers and prefigured the ending of the Cold War. In this speech, President Reagan addressed the United Nations about a range of topics—the continued Soviet presence in Afghanistan, American foreign policy goals in Central America, and the Middle East peace process. He also discussed the INF treaty and the potential changes in American-Soviet relations. How does he characterize these changes? How does he tie his vision for the future to the stated domestic and foreign policy aims of his administration? What elements of the Cold War remain in his language and rhetoric?

I've come here today to map out for you my own vision of the world's future, one, I believe, that in its essential elements is shared by all Americans. And I hope those who see things differently will not mind if I say that we in the United States believe that the place to look first for

Excerpted from Ronald Reagan, "Address to the 42nd Session of the United Nations General Assembly," September 21, 1987, Ronald Reagan Presidential Library, www.reagan.utexas.edu/archives/speeches/1987/092187b.htm.

shape of the future is not in continental masses and sea lanes, although geography is, obviously, of great importance. Neither is it in national reserves of blood and iron or, on the other hand, of money and industrial capacity, although military and economic strength are also, of course, crucial. We begin with something that is far simpler and yet far more profound: the human heart.

All over the world today, the yearnings of the human heart are redirecting the course of international affairs, putting the lie to the myth of materialism and historical determinism. We have only to open our eyes to see the simple aspirations of ordinary people writ large on the record of our times. . . .

Those who advocate statist solutions to development should take note: The free market is the other path to development and the one true path. And unlike many other paths, it leads somewhere. It works. So, this is where I believe we can find the map to the world's future: in the hearts of ordinary people, in their hopes for themselves and their children, in their prayers as they lay themselves and their families to rest each night. These simple people are the giants of the Earth, the true builders of the world and shapers of the centuries to come. And if indeed they triumph, as I believe they will, we will at last know a world of peace and freedom, opportunity and hope, and, yes, of democracy—a world in which the spirit of mankind at last conquers the old, familiar enemies of famine, disease, tyranny, and war.

This is my vision—America's vision. I recognize that some governments represented in this hall have other ideas. Some do not believe in democracy or in political, economic, or religious freedom. Some believe in dictatorship, whether by one man, one party, one class, one race, or one vanguard. To those governments I would only say that the price of oppression is clear. Your economies will fall farther and farther behind. Your people will become more restless. Isn't it better to listen to the people's hopes now rather than their curses later?

And yet despite our differences, there is one common hope that brought us all to make this common pilgrimage: the hope that mankind will one day beat its swords into plowshares, the hope of peace. . . .

We're heartened by new prospects for improvement in East-West and particularly U.S.-Soviet relations. Last week Soviet Foreign Minister [Eduard] Shevardnadze visited Washington for talks with me and with the Secretary of State, [George] Shultz. We discussed the full range of issues, including my longstanding efforts to achieve, for the first time, deep reductions in U.S. and Soviet nuclear arms. It was 6 years ago, for example, that I proposed the zero-option for U.S. and Soviet longer range, intermediate-range nuclear missiles. I'm pleased that we have now agreed in principle to a truly historic treaty that will eliminate an entire class of U.S. and Soviet nuclear weapons. We also agreed to intensify our diplomatic efforts in all areas of mutual interest. Toward that end, Secretary Shultz and the Foreign Minister will meet again a month from now in Moscow, and I will meet again with General Secretary [Mikhail] Gorbachev later this fall.

We continue to have our differences and probably always will. But that puts a special responsibility on us to find ways—realistic ways—to bring greater stability to our competition and to show the world a constructive example of the value of communication and of the possibility of peaceful solutions to political problems. And here let me add that we seek, through our Strategic Defense Initiative, to find a way to keep peace through relying on defense, not offense, for deterrence and for eventually rendering ballistic missiles obsolete. SDI has greatly enhanced the prospects for real arms reduction. It is a crucial part of our efforts to ensure a safer world and a more stable strategic balance.

We will continue to pursue the goal of arms reduction. . . .

We look forward to a time when things we now regard as sources of friction and even danger can become examples of cooperation between ourselves and the Soviet Union. For instance, I have proposed a collaboration to reduce the barriers between East and West in Berlin and, more broadly, in Europe as a whole. Let us work together for a Europe in which force of the threat—or, force, whether in the form of walls or of guns, is no longer an obstacle to free choice by individuals and whole nations. I have also called for more openness in the flow of information from the Soviet Union about its military forces, policies, and programs so that our negotiations about arms reductions can proceed with greater confidence.

We hear much about changes in the Soviet Union. We're intensely interested in these changes. We hear the word *glasnost*, which is translated as "openness" in English. "Openness" is a broad term. It means the free, unfettered flow of information, ideas, and people. It means political and intellectual liberty in all its dimensions. We hope, for the sake of the peoples of the U.S.S.R., that such changes will come. And we hope, for the sake of peace, that it will include a foreign policy that respects the freedom and independence of other peoples. . . .

I have spoken today of a vision and the obstacles to its realization. More than a century ago a young Frenchman, Alexis de Tocqueville, visited America. After that visit he predicted that the two great powers of the future world would be, on one hand, the United States, which would be built, as he said, "by the plowshare," and, on the other, Russia, which would go forward, again, as he said, "by the sword." Yet need it be so? Cannot swords be turned to plowshares? Can we and all nations not live in peace? In our obsession with antagonisms of the moment, we often forget how much unites all the members of humanity. Perhaps we need some outside, universal threat to make us recognize this common bond. I occasionally think how quickly our differences worldwide would vanish if we were facing an alien threat from outside this world. And yet, I ask you, is not an alien force already among us? What could be more alien to the universal aspirations of our peoples than war and the threat of war?

Two centuries ago, in a hall much smaller than this one, in Philadelphia, Americans met to draft a Constitution. In the course of their debates, one of them said that the new government, if it was to rise high, must be built on the broadest base: the will and consent of the people. And so it was, and so it has been.

My message today is that the dreams of ordinary people reach to astonishing heights. If we diplomatic pilgrims are to achieve equal altitudes, we must build all we do on the full breadth of humanity's will and consent and the full expanse of the human heart. Thank you, and God bless you all.

EDITORS OF THE *BULLETIN OF THE ATOMIC SCIENTISTS*

"A NEW ERA," 1991

In 1947 the editors of the *Bulletin of the Atomic Scientists*—a periodical founded by several Manhattan Project scientists and intended to inform the public about developments in nuclear technology—introduced the Doomsday Clock. The clock symbolically represented the threat of a global nuclear war. The clock moved to two minutes before midnight after American and Soviet testing of thermonuclear weapons, to twelve minutes after the signing of the Partial Test Ban Treaty in 1963, and to three minutes in 1984 in response to the renewed tensions between the superpowers. As the editors of the *Bulletin* explain in this essay, the end of the Cold War sparked a new optimism about the risk of a nuclear confrontation. That said, what do the authors identify as continuing concerns? How do they see nuclear energy interacting with other elements of human security?

. . . .

With this issue, the *Bulletin* resets the *Bulletin* Clock from 10 to 17 minutes until midnight. The clock is in a new region because we feel the world has entered a new era. Never before has the Board of Directors moved the minute hand so far at one time. Conceived at the dawn of the Cold War, the Clock was designed with a 15-minute range. John A.

"A New Era," editors of the *Bulletin of the Atomic Scientists* 47, no. 10 (December 1991): 3. Reprinted by permission of the publisher, Taylor & Francis Ltd, www.tandfonline.com.

Simpson, one of the *Bulletin*'s founders, says that a 15-minute scale was all anyone thought would be needed in their lifetimes. The present move was not easily agreed upon. Board members initially expressed divergent views as did some of the sponsors of the *Bulletin*. But on balance a consensus was reached reflecting a conviction that the world was changing in fundamental and positive ways.

Foremost are the developments in East-West relations. The Strategic Arms Limitation Treaty was revived and promptly signed. Shortly thereafter hardliners in the Soviet Union mounted a coup that quickly failed and, among other results, significantly reduced what might have been years of negotiations between our two countries. Then, on September 27, President George Bush announced the withdrawal of thousands of tactical weapons. Many strategic missiles were taken off hair-trigger alert, as was the B-52 bomber fleet. On October 5, President Mikhail Gorbachev announced similar initiatives and upped the ante by indicating that the Soviet Union would suspend nuclear testing. We hope the United States will have a positive response.

The Cold War is over. The 40-year-long East-West nuclear arms race has ended. The world has clearly entered a new post–Cold War era. The illusion that tens of thousands of nuclear weapons are a guarantor of national security has been stripped away. In the context of a disintegrating Soviet Union, large nuclear arsenals are even more clearly seen as a liability, a yardstick of insecurity.

But the world is still a dangerous place. The START [Strategic Arms Reduction Talks] agreement, which mandates reductions in the number of strategic nuclear weapons, does not initiate a process for retiring the warheads and converting the highly enriched uranium and plutonium into a form not usable in weapons. Indeed it does not require the destruction of warheads, but it does permit modernization of nuclear forces. Agreement by both sides to cease testing would restrict this.

While the failed coup and subsequent extraordinary changes in the "former" Soviet Union are major reasons for optimism, the whole world is properly concerned with the risks of national disintegration and the security of the 27,000 nuclear weapons during this critical period when the republics and the central government strive to create a new reliable relationship despite great difficulties.

We recognize that nuclear weapons remain; even the modest agreed-upon reductions in their numbers will not be achieved overnight. Other nations have such weapons and more aspire to have them—impelled, in part, by the failure of the nuclear superpowers to fulfill the terms of the Non-Proliferation Treaty, thereby rendering themselves incapable of invoking the treaty for others.

The *Bulletin* will continue to keep a wary eye on the nuclear weapons situation, but we will also focus on new concepts of security that are of critical importance, and we will more intensely address the militarization of our society and of our national policy and that of other nations. We will be concerned about the sale of weapons and of facilities used in manufacturing them, and with all aspects of manufacture and distribution that make up the world's trillion-dollar investment in arms and armies.

Far deeper reductions in nuclear weapons stockpiles must follow. The *Bulletin* has long reported on the destructiveness of seeking military solutions to the world's ills, and it will continue to do so. It has long reported on the economic distortions and human misery caused when the nations of the world pour vast sums of money and intellectual capital into weaponry—as they still do even though the paralyzing shadow of nuclear apocalypse has faded. We will aggressively examine the opportunities presented by the end of the East-West nuclear arms race.

We are encouraged by East-West agreements to conduct on-site inspections, and by the improvement of technologies for verification. The experience of the Gulf War offers many lessons, one of the most significant being the increasing role for the United Nations in resolving conflict and blocking aggression. The *Bulletin* must and will continue to address these issues as well as the proliferation of weapons, the conditions of nuclear power, and environmental concerns that threaten in the long run the well-being—indeed, the security—of all peoples.

We believe that presidents Bush and Gorbachev have guided their respective nations to a historic intersection of mutual interests. Continuing boldness and imagination are called for. Men and women throughout the world must vigorously challenge the bankrupt paradigms of militarism if we are to achieve a new world order. The setting of the *Bulletin* Clock reflects our optimism that we are entering a new era.

EPILOGUE

THE NUCLEAR PRESENT

Many of the geopolitical, economic, and environmental factors that shaped American encounters with nuclear energy have changed since the end of the Cold War; many others have remained the same. The global battle against communism has been replaced by the War on Terror. The cold calculus of deterrence and "mutually assured destruction" no longer serve to keep the threat of nuclear war at bay. The threat of climate change has caused many people—including some environmentalists who once opposed nuclear power—to reconsider nuclear as a low-carbon energy source. Still others argue that we cannot proceed with nuclear power until we resolve the radioactive waste dilemma. Americans' relationships to nuclear energy are as shifting and uncertain as they have been at any time since the dawn of the Atomic Age. The documents in the epilogue illustrate some of the questions that Americans have been asking about nuclear energy since the end of the Cold War. In some cases, the same questions have persisted since the beginning of the nuclear age; other questions are new to the context of the twenty-first century. But the nuclear paradox remains: nuclear energy still seems to offer both the promise of salvation and the risk of ruin.

The nuclear threat seems distant to Americans born after the end of the Cold War, and most people today give little thought to the prospect of nuclear war or a reactor meltdown. Local places contaminated by their nuclear histories or targeted as radioactive storage facilities, however, still wrestle with these issues. From a post–Cold War perspective, the contamination of these places emerges as a much more significant environmental legacy than the global concerns about nuclear winter or the postattack environment that motivated the antinuclear movement.

The earthquake, tsunami, and meltdown at the Fukushima Daiichi plant in Japan in 2011 are reminders that nuclear energy still comes with significant risk to human and environmental health.

The nuclear past parallels modern debates about nuclear energy. As you read through the documents in this section of the book, consider these parallels between past encounters with nuclear energy and current debates about its use and management. How have the practice, use, and politicization of science changed over time? How do we deal with the differentiated burdens of risk created by radioactive waste? How does nuclear energy still pose challenges to civic questions such as community identity and decision-making? Many of the questions raised by encounters with nuclear energy are being asked again, but in a new context framed by concerns about climate change. How has scientific uncertainty shaped the climate change debate? What represents "progress" in the face of climate change? Who benefits from the production of greenhouse gasses, and who faces the risk of harm from environmental change? In short, climate change is forcing us to reconsider our ideas about human relations with nature, the shape of civil society, and the meaning of progress. It therefore makes sense to look at the ways that nuclear energy forced Americans to wrestle with these same issues.

DAVID ALBRIGHT, KATHRYN BUEHLER,
AND HOLLY HIGGINS

"BIN LADEN AND THE BOMB," 2002

The end of the Cold War marked a radical change in the assessment of risk and nuclear weapons. As the Soviet Union disintegrated, grave questions arose about the safeguarding of its nuclear stockpiles. The attack on New York on September 11, 2001, raised a host of new concerns about national security and nuclear security. In this essay United Nations nuclear weapons inspector David Albright addressed the chilling prospect of whether or not the terrorist organization Al Qaeda could obtain a nuclear weapon. How had the threat of nuclear weapons changed with the end of the Cold War? How does this perspective compare to assessments of nuclear threats during the Cold War?

. . . .

Are Osama bin Laden and the Al Qaeda network shopping for nuclear weapons—or do they already have them? Is it possible that today's terrorists could acquire such weapons—and use them? Answering these questions has taken on an added urgency in the wake of the September 11 attacks. Al Qaeda has sought nuclear weapons for years. Last year, the CIA intercepted a cryptic message from an Al Qaeda member who boasted that Osama bin Laden was planning to carry out a "Hiroshima" against America.

David Albright, Kathryn Buehler, and Holly Higgins, "Bin Laden and the Bomb," *Bulletin of the Atomic Scientists* 58, no. 1 (January–February 2002): 22–23.

Nuclear weapons in the hands of terrorists is a frightening prospect. A surface detonation in a major U.S. city of a five-kiloton nuclear bomb—one-third the size of the Hiroshima blast—would destroy most buildings within a several-block radius. Many within about a mile from ground zero would receive severe radiation and burn injuries.

Considering the stakes, everyone wants 100 percent certainty that terrorist groups do not have nuclear weapons. But absolute certainty is impossible, so governments must exert extraordinary efforts to ensure that terrorists never acquire such weapons.

Following extensive analysis of open source information and interviews with knowledgeable officials, the Institute for Science and International Security found no credible evidence that either bin Laden or Al Qaeda possesses nuclear weapons or sufficient fissile material to make them. However, if Al Qaeda obtained enough plutonium or highly enriched uranium, we believe it is capable of building a crude nuclear explosive, despite several difficult steps. We cannot say absolutely whether Al Qaeda possesses fissile material, but to our knowledge no evidence of possession has surfaced.

This uncertainty reflects several factors. We know of previous attempts—all unsuccessful—by Al Qaeda agents to buy highly enriched uranium in the mid-1990s in Africa, Europe, and Russia. Bin Laden has loudly proclaimed his desire for nuclear capability, and on November 9, he told a Pakistani journalist that he already has nuclear weapons. U.S. intelligence officials reportedly believe that bin Laden is actively seeking nuclear weapons, but they doubt his claim that he possesses any. . . .

If they had a secret, fixed base in Afghanistan, over the last several years Al Qaeda and its Taliban allies could have made significant progress on nuclear research. Such a base would be beneficial to nuclear weaponization activities, particularly in overcoming engineering and other practical steps in building a weapon.

If Al Qaeda were to build nuclear weapons, it would likely build relatively crude, massive nuclear explosives, deliverable by ships, trucks, or private planes. Stopping such an attack would be extremely difficult.

WHAT NEEDS TO BE DONE

The United States and its allies must continue to scour Afghanistan, searching for evidence of Taliban or Al Qaeda nuclear activity, trying to identify the effort's scope, origin, timing, and purpose. It is critical to determine what the Taliban or Al Qaeda have already accomplished; to identify and destroy any nuclear equipment, materials, or facilities; and to gather intelligence about Al Qaeda and its allies' nuclear activities outside of Afghanistan.

Another priority should be locating any scientists, officials, or technicians involved in Al Qaeda or Taliban nuclear efforts, and encouraging them (through incentives or threat of jail time) to talk. Investigators should focus especially on whether nuclear or nuclear-related items were obtained from overseas, and if so, who the suppliers were.

Efforts to find international Al Qaeda "sleeper cells" that may be working to master nuclear crafts should be accelerated. Any Al Qaeda nuclear specialists or bright scientists or technicians willing to learn about nuclear weapons remain a threat.

All nuclear weapons and fissile material must be better secured to minimize the chance that terrorists will someday get their hands on nuclear weapons. The problem of poorly protected stockpiles is most acute in Russia, which in 2000 still possessed an estimated 1,150 metric tons of weapon-grade plutonium and highly enriched uranium. Only a relatively small amount of fissile material—from a few to tens of kilograms—is needed to make a nuclear explosive. Better control, accounting, and protection is needed to ensure that a terrorist group cannot secretly obtain any of this material.

More effective coordination between key governments is essential to guarantee that terrorists never acquire nuclear weapons. A well-organized terrorist group could try to develop its nuclear capabilities in many countries at once. Terrorist groups must be aggressively pursued, and individual governments must work to prevent cells from operating within their countries on nuclear weapons activities. . . .

Governments should establish an international group, advised or staffed by nuclear weapons experts, with the authority to investigate terrorists' nuclear activities and to coordinate with national law enforcement and intelligence agencies. An international group, even if loosely defined, could also help educate the public about the threat of nuclear terrorism, and perhaps even sound an early alarm to which national governments or the U.N. Security Council could respond collectively to thwart a nuclear attack.

Preventing terrorists from striking with nuclear weapons will not be easy, but it will be worth the effort. Armed with nuclear weapons, terrorists could fracture civilization.

ALLISON M. MACFARLANE

"YUCCA MOUNTAIN AND HIGH-LEVEL NUCLEAR WASTE DISPOSAL," 2006

Although Congress had selected Yucca Mountain in Nevada as the site for a geologic repository for the storage of radioactive wastes in 1987, the decision remained highly controversial, bogged down by opposition from the state of Nevada, questions about site safety, and bureaucratic delays. Meanwhile, spent fuel accumulated at commercial reactors around the country, increasing the need to find a solution to the waste dilemma. In the document excerpted below, geologist Allison Macfarlane—one of the nation's foremost scientific experts on waste storage and, later, the chair of the Nuclear Regulatory Commission (NRC)—testifies about the ongoing uncertainty at Yucca Mountain. What concerns does she raise about the science of geologic siting and the process used to verify the safety of Yucca Mountain? Why does she believe that the search for a geologic repository must continue? How might her concerns about probabilistic models and their impact on decision-making parallel debates about the science of global climate change?

. . . .

Let me begin by emphasizing that in my expert opinion, the best solution to the problem of high-level nuclear waste remains a geologic repository. On this issue all countries with nuclear energy programs

---

Excerpted from Allison M. Macfarlane, "Yucca Mountain and High-Level Nuclear Waste Disposal," in Senate Committee on Environment and Public Works, *Status of the Yucca Mountain Project: Hearing before the Committee on Environment and Public Works*, 109th Congress, 2d session, March 1, 2006.

are in agreement. . . . In light of the push for more nuclear power in the U.S., . . . it is highly likely that multiple Yucca Mountain–type repositories will be necessary. Therefore, it is imperative that we continue to work towards a solution to the problem of high-level nuclear waste.

Some policymakers have suggested that long-term above-ground storage of spent fuel is a better solution to the current problem. Their idea is to wait until a better alternative to geologic disposal is discovered. Interim storage is just that—an interim, temporary solution. Interim storage is fine for 100 years, but longer than that one cannot be assured the containers would prevent radioactivity from entering the environment. In the unlikely case that societal control is lost over the interim storage site or technological advance cannot provide a better alternative to geologic disposal in the next 100 years, interim storage fails its task and exposes future generations to radioactivity. Thus, I would argue for continued work on geologic repository disposal of high-level nuclear waste. . . .

UNCERTAINTY AND YUCCA MOUNTAIN

There are many uncertainties associated with trying to understand the behavior of a high-level nuclear waste repository thousands or hundreds of thousands of years into the future. One question we need to ask in siting a repository is whether the earth system is well enough understood to make predictive models of a repository far into the future? Is it possible to verify or validate these models? If not, then can one site a repository?

The DOE [Department of Energy] has argued that it has characterized all the relevant "features, events, and processes" at Yucca Mountain. I will argue that from my geologist's viewpoint that the DOE cannot know all the features, events, and processes it needs to describe the repository system because the repository is an evolving system whose basic thermodynamic and kinetic features are still not known. . . .

Perhaps our current Defense Secretary, Donald Rumsfeld, put it best by noting in a 2002 press briefing, "There are known knowns; there are things we know we know. We also know there are known unknowns; that is to say we know there are some things we do not know. But there are also unknown unknowns—the ones we don't know we don't know."

What we don't know we don't know could prove to be very important in the behavior of a geologic repository.

The DOE has attempted to predict the behavior of the Yucca Mountain repository over time using a complex computer modeling method called probabilistic performance assessment. The performance assessment of the Yucca Mountain repository is made up of numerous submodels of systems that will affect repository behavior such as the climate, the unsaturated zone, the waste package, etc. The DOE has stated that it has validated these models by the use of laboratory tests, in situ tests, and field tests.

From the perspective of an earth scientist, it is not possible to validate or verify models of earth systems. This is because earth systems are by definition open systems, accessible to exchanges of matter and energy. As a result, in open systems, it is not possible to know all the potential processes or input parameters that might affect the system. The Yucca Mountain repository is one of those open systems, and therefore it is not possible to legitimately validate the performance assessment model.

Models of earth systems cannot be validated or verified by comparison to laboratory, in situ or field data for two reasons. First, the data may have errors in it that while small now, over time may result in a large deviation from actual behavior. Second, though model results may predict current behavior, over time the geologic system will change in unpredictable ways, and therefore it is not possible to predict future conditions.

The terms "validate" and "verify" powerfully signify the truth of model results, suggesting that the model is an accurate representation of future behavior of the system. These terms are used to convince policymakers of the truth of the model results, though in actuality, the models cannot be validated or verified.

More disturbing is a practice, perhaps an unconscious one, in which experts present model results as if they were actual data. Secretary of Energy [Spencer] Abraham was guilty of such practice when he stated, "The amount of water that eventually reaches the repository level at any point in time is very small. . . ." We have not and cannot measure the amount of water that will reach the repository at any time in the future, but the DOE generated a model of the amount of water that might reach the repository, which provided the results stated by the Secretary. These are not *facts*, but instead unvalidatable *model results*. . . .

Why all the emphasis on performance assessment? The results of probabilistic performance assessment will be used by both the DOE and the NRC to determine the suitability of the Yucca Mountain site. They are forced to use these complex models for two reasons. First, there is only one site to evaluate, so it cannot be evaluated in a relative sense, as was the plan in the 1982 Nuclear Waste Policy Act (NWPA). The 1987 Nuclear Waste Policy Amendments Act changed this strategy by allowing the characterization of only a single site, Yucca Mountain. Thus, the DOE and NRC needed to develop a method to evaluate the site in an absolute sense. They decided by the early 1990s that performance assessment modeling was advanced enough to apply to a geologic repository. . . .

WHAT SHOULD WE DO?

Given the strict limits placed by the EPA [Environmental Protection Agency] on the DOE and NRC's ability to evaluate the Yucca Mountain site and their inability to determine whether the numbers produced by the performance assessment models are valid, how should the United States determine the suitability of a repository site? I would like to make the following suggestions. . . .

. . . [W]ork must continue on the Yucca Mountain site to determine whether it will be suitable as a geologic repository. To supplement the performance assessment, which would be useful only for short time periods (on the order of centuries), a comparative analysis can be adopted similar to that proposed in the 1982 NWPA. As it is not practical or pragmatic to select other U.S. sites and begin in-depth characterization for the purposes of comparison with Yucca Mountain, I suggest an alternative method. A large body of data exists for a number of investigated repository sites around the world. I suggest that this dataset be used for comparative purposes with Yucca Mountain. . . .

. . . [I]f Yucca Mountain is found lacking in comparison to the above-listed sites, the DOE and the NRC may decide that it is not appropriate for use as a geologic repository. In this case, Congress would need to revisit repository siting and issue new legislation that allows the DOE to search for and establish new sites. It is highly likely that Congress will have to address this issue in the next ten years even if Yucca Mountain is approved by the NRC because it will not be able to contain all the waste produced in this country. In the United States, we are fortunate to have a large country with many geologically appropriate locations for a nuclear waste repository that have arguably simpler geology than Yucca Mountain. For a repository to succeed, the process must be fair and perceived to be fair by all participants.

A large amount of high-level nuclear waste already exists in the United States and requires disposal. This problem deserves rapid and focused attention. It is resolvable, but requires a delicate balance of technical prowess and fair and just policy-making. For the betterment of our environment, it is within our grasp to solve this problem.

OREGON DEPARTMENT OF ENERGY

HANFORD CLEANUP: THE FIRST 20 YEARS, 2009

As the Cold War ended, most Americans have given less and less thought to the risks posed by nuclear weapons. Some communities do not have this luxury. Rocky Flats in Colorado, the Nevada Test Site, Los Alamos in New Mexico, and many other locations where the U.S. government researched, tested, and constructed nuclear weapons deal with the legacy of radioactive contamination on a daily basis. Many of these places built their economies and civic identities out of their position on the front lines of the Cold War. For these communities the environmental costs of nuclear weapons production now loom larger than the benefits provided by the nuclear shield. More than any other place, the Hanford Site in western Washington has witnessed the costs and challenges of contamination from Cold War weapons production; today it is the nation's most contaminated landscape and the location of the world's largest environmental remediation effort. How do the participants involved in the cleanup describe the challenges they have faced? What lessons do they suggest should be learned from Hanford's encounters with nuclear energy? What are the legacies of the nuclear consensus, from the perspective of places such as Hanford?

• •

May 15, 2009, marked the twentieth anniversary of the signing of the Hanford Federal Facility Agreement and Consent Order, most often

---

Excerpted from Oregon Department of Energy, *Hanford Cleanup: The First 20 Years* (Salem: Oregon Department of Energy, 2009), i–ii, v–xi.

referred to as the Tri-Party Agreement. The signing of the Tri-Party Agreement marked the formal beginning of cleanup of the Hanford nuclear site in Washington state. The agreement, signed by the U.S. Department of Energy (DOE), the U.S. Environmental Protection Agency, and the Washington Department of Ecology, established a 30-year timetable for cleaning up Hanford's toxic wastes.

The 586 square mile Hanford Site was home to the world's first plutonium production facilities. The processes used at Hanford to create plutonium generated tremendous amounts of radioactive and chemically hazardous waste. Plutonium production ended at Hanford in 1988. Since 1989, the focus has been on environmental cleanup.

By most estimates, the Hanford cleanup is just a third of the way complete. Amendments to the Tri-Party Agreement have so far extended the end of cleanup by another decade and may yet extend it by two additional decades. Cleanup will likely continue well into the 2040s and possibly beyond.

This continuing extension perhaps best illustrates the massive extent of contamination resulting from 45 years of plutonium production for America's nuclear weapons program. It is also a reflection on the waste management practices used during much of that time.

There is no doubt there has been considerable progress at Hanford. The threat to the public and the environment has been dramatically reduced since those days when Hanford began the awkward transition from production to cleanup. Along the way, those involved overcame some tremendously difficult and unique technical challenges.

But the cleanup is also not nearly as far along as any of us expected or would like to have seen. The Hanford cleanup has been more difficult, more costly (about $30 billion so far), and has taken much longer than anyone envisioned at the start. The remaining challenges will require significant funds, technical ingenuity, and dogged determination to see the cleanup through to completion. . . .

The Columbia River flows through the Hanford Site then continues downstream past prime Oregon farmlands, fisheries and recreation areas. The Hanford Site includes the Hanford Reach, a major spawning area for Chinook salmon and steelhead. Radioactive wastes transported to and from Hanford travel on at least 200 miles of Oregon highways. Portions of two Oregon counties are within the 50 mile nuclear emergency planning radius of Hanford. People in these areas could be at risk in the event of a major Hanford accident. A quality cleanup at Hanford is vital to protect Oregon's citizens and our natural resources. . . .

We believe the history of Hanford cleanup offers us lessons for the present and for the future and is well worth documenting. We have already seen that assumptions made during the operating years about the finality of waste disposal have in many cases proven to be very wrong. Considerable effort has gone into digging up many old burial grounds and disposal areas that were thought at the time to be safe and permanent disposal places. We hope that decisions and actions that have been made during these past 20 years are protective and durable.

The biggest lesson may be one that has been verbalized many times, yet often ignored—by Congress, by DOE, by regulators, by many of us. That lesson is that there are few quick and inexpensive solutions at Hanford. The extent of the contamination or the complexity of the solutions—or both—generally prevent speedy or cheap resolution. . . .

REFLECTIONS ON THE BEGINNING OF THE HANFORD CLEANUP

In the spring and early summer of 2009, the Oregon Department of Ecology contacted a number of people who were involved at the beginning of Hanford cleanup. We asked them to reflect back on some of their most vivid memories of that period and their expectations at that time for the cleanup to come.

Through telephone calls and e-mails, we heard from the following:

- James Watkins, Secretary of Energy, 1989–1993.
- W. Henson Moore, Deputy Secretary of Energy, 1989–1992. . . .
- Roger Stanley, former Lead Tri-Party Agreement Negotiator for the Washington Department of Ecology.
- Randy Smith, former Lead Tri-Party Agreement Negotiator for the U.S. Environmental Protection Agency (EPA). . . .
- Bill Dixon, former Administrator of Nuclear Safety and Energy Facilities for the Oregon Department of Energy. . . .

Bill Dixon: "(Resistance to cleanup) began to change in about 1986 when . . . [it] became evident to many more people that this was one of the most significant risks in the Northwest that must be faced soon. Otherwise, future generations would curse us for allowing the lifeblood of the Northwest—the Columbia River—to become contaminated with Hanford's wastes. These concerns spread like wildfire throughout the Northwest, and largely due to political pressure on DOE from Washington and Oregon, the DOE leaders in Washington, D.C. agreed to face the issue." . . .

W. Henson Moore: "It was a culture change and we were pounded by people within the Defense department. We were pounded by people within the Department of Energy. We were pounded by contractors, who were essentially resisting this change from production to cleanup. We were being pushed on all fronts. Because quite frankly, a culture had been developed in building the nuclear weapons for the nation's defense. And that came first. It came first before everything else. The environment wasn't even a consideration. Health and safety was less of a consideration. The consideration was succeed at all cost. The nation's defense depended upon it." . . .

Randy Smith: "At the time the Tri-Party Agreement was initially negotiated, the toughest negotiations were over what could be done and would be done about the tank wastes. The uncertainties about what was in the tanks, how to sample them to better understand their contents, and how to develop feasible processes to remove and treat the wastes were enormous. The initial Tri-Party Agreement was quite sketchy about the details of what ought to be done. It was clear to all of us, and to any observer, that the tank wastes . . . would be the thorniest problem. I remember thinking that there was no way that these problems would be dealt with within the careers of those then working at Hanford or for Ecology or EPA. Our children, or maybe our grandchildren's generation would still be working on this remediation. I didn't think of that as pessimism, just realism." . . .

Roger Stanley: "We were unaware of the extent of contamination from tank leaks; that DOE's tanks would continue to fail; or of the volumes of tank waste and contaminants associated with the many miles of pipelines and sumps that run between the tanks themselves. Though I was never one to press for anything close to pristine conditions upon closure, it was always somewhat of a disappointment that despite the fact that a teacup's worth of tank waste will remain extremely hazardous virtually forever, the sheer magnitude and cost of the project causes people to lower their sights."

W. Henson Moore: "There was a toxic mixture of materials in those tanks at Hanford that was new to mankind. The only time you would have something like that—us and probably the Soviets had it—is because of the nuclear weapons program. And you would have this leftover stuff—toxic waste—you'd be mixing and putting in these tanks. And I just remember that nobody could answer satisfactorily—for me as a neophyte and for a non-scientist—what was really going on in the tanks and what we're going to have to do with them. We just knew that, Lord help us, if one of those things ever blew up, or if it ever collapsed for some reason and allowed what was contained in the tanks to run free. Because it was just a real toxic mix of stuff."

Bill Dixon: "When we got into the job, we found that it was much harder than any of us had imagined. We didn't know how much waste and contamination existed, where it all was, or what was in it. Most technologies and work practices needed to clean up the waste did not exist. We had not addressed how workers could perform this work safely. Original cost estimates were only a small fraction of what was needed. And, the existing environmental laws, regulations and processes were nearly unworkable for a cleanup this massive. As we struggled with these complications, and many more, schedules began to stretch and costs began to skyrocket." . . .

James Watkins: "Hanford waste is the example of what goes sour when you don't pay attention to all of these other things that get into the ecology, health, safety of human beings, safety of operations, the top level people from Washington on down knowing what is going on out there in the labs and in the burial grounds and so forth. . . . We had to get control of these things, it was not under control. And that was one of the things that I found and I told the President, 'You want me to clean this place up, you better help me.' Because a lot of people aren't going to like it, see us putting money into things we haven't put it before."

MARK Z. JACOBSON

"NUCLEAR POWER IS TOO RISKY," 2010

As the scientific consensus over anthropogenic climate change grew more certain and the expected impacts of climate change seemed more catastrophic, the debate over nuclear energy changed. Industry executives recognized a new opening for commercial nuclear power. Many environmental activists also saw nuclear energy as one of the most promising responses to the threat of climate change. Others remained unconvinced, however, pointing to the industry's ongoing struggles with safety, waste disposal, and a broken regulatory process. In this essay, Stanford University professor of environmental engineering Mark Jacobson lays out the case against nuclear energy as a low-carbon energy source. How does he come to this conclusion? Does he root his argument in economic, environmental, or social concerns? How does nuclear history inform his perspective?

....

If our nation wants to reduce global warming, air pollution and energy instability, we should invest only in the best energy options. Nuclear energy isn't one of them.

Every dollar spent on nuclear is one less dollar spent on clean renewable energy and one more dollar spent on making the world a

---

Mark Z. Jacobson, "Nuclear Power Is Too Risky," February 22, 2010, www.cnn.com/2010/OPINION/02/22/jacobson.nuclear.power.con/. Courtesy of Mark Z. Jacobson.

comparatively dirtier and a more dangerous place, because nuclear power and nuclear weapons go hand in hand.

In the November [2009] issue of *Scientific American*, my colleague Mark DeLucchi of the University of California–Davis and I laid out a plan to power the world with nothing but wind, water and sun. After considering the best available technologies, we decided that a combination of wind, concentrated solar, geothermal, photovoltaics, tidal, wave and hydroelectric energy could more than meet all the planet's energy needs, particularly if all the world's vehicles could be run on electric batteries and hydrogen fuel cells.

We rejected nuclear for several reasons. First, it's not carbon-free, no matter what the advocates tell you. Vast amounts of fossil fuels must be burned to mine, transport and enrich uranium and to build the nuclear plant. And all that dirty power will be released during the 10 to 19 years that it takes to plan and build a nuclear plant. (A wind farm typically takes two to five years.)

The on-the-ground footprint of nuclear power, through its plants and uranium mines, is about 1,000 times larger than it is for wind. Wind turbines are merely poles in the ground—with lots of space between them that can be farmed, ranched or left open—or poles in the ocean. Geothermal energy also has a much smaller footprint than nuclear; solar only slightly more. But while geothermal, solar and wind are safe, nuclear is not.

For nuclear to meet all the world's energy needs today—12.5 terawatts (1 terawatt = 1 trillion watts)—more than 17,000 nuclear plants would be needed. Even if nuclear were only 5 percent of the solution, most countries would have nuclear plants.

What's worse, the nuclear industry wants to reprocess waste to obtain more energy from increasingly scarce uranium. But this only produces more weapons-grade uranium and plutonium.

A global push toward nuclear energy would mean that uranium enrichment—and efforts at nuclear weapons development—would certainly grow throughout the world.

Nuclear proponents argue that not enough clean renewables exist to power the world. However, part of our work at Stanford University has been to map world renewable energy resources. Enough wind and solar exist in high-wind and sunny locations over land to power the world for all purposes multiple times over. There is no shortage.

Nuclear proponents also argue that nuclear energy production is constant, unlike fickle winds and sunshine. But worldwide, nuclear plants are down 15 percent of the time, and when a plant goes down, so does a large fraction of the grid. Connecting wind farms over large areas through transmission lines smoothes power supply. Combining geothermal with wind (whose power potential often peaks at night) and solar (which peaks by day), and using hydroelectricity to fill in gaps, would almost always match demand.

Converting to electric vehicles and using smart charging practices would also help to match supply with demand. So would storing energy (with concentrated solar) and giving people incentives to reduce demand. It is not rocket science to match power demand. It merely requires thinking out of the box.

Finally, the costs of land-based wind, geothermal and hydroelectricity are competitive with conventional new sources of electricity; costs of solar and wind over the ocean are higher but declining. Costs of nuclear have historically been underestimated.

In sum, if we invest in nuclear versus true renewables, you can bet that the glaciers and polar ice caps will keep melting while we wait, and wait, for the nuclear age to arrive. We will also guarantee a riskier future for us all.

There is no need for nuclear. The world can be powered by wind, water and sun alone.

PRESIDENT'S BLUE RIBBON COMMISSION ON AMERICA'S NUCLEAR FUTURE

REPORT TO THE SECRETARY OF ENERGY, 2012

In 2010, President Barack Obama created the Blue Ribbon Commission on America's Nuclear Future, charging its fifteen members—a mix of politicians, academics, environmental activists, and energy industry executives—to review policies and options for managing the country's ever-increasing supply of radioactive waste. After more than six decades of false starts, technical failures, and public protests, the United States still has no workable solution to the question of what to do with spent fuel and other radioactive waste. The report identified the need for a consent-based approach to finding a geologic repository for the storage of waste. How and why do the commissioners come to this conclusion? Why do they see radioactive waste as such an urgent problem, and how do they propose to move the nation past the impasse that has prevented solutions in the past? In what ways does the report reflect—or ignore—the long history of America's interactions with nuclear energy?

America's nuclear waste management program is at an impasse.

The Obama Administration's decision to halt work on a repository at Yucca Mountain in Nevada is but the latest indicator of a policy that

Excerpted from President's Blue Ribbon Commission on America's Nuclear Future, *Report to the Secretary of Energy*, 2012, http://energy.gov/ne/downloads/blue-ribbon-commission-americas-nuclear-future-report-secretary-energy, pp. vi–ix, xv.

has been troubled for decades and has now all but completely broken down. The approach laid out under the 1987 Amendments to the Nuclear Waste Policy Act (NWPA)—which tied the entire U.S. high-level waste management program to the fate of the Yucca Mountain site—has not worked to produce a timely solution for dealing with the nation's most hazardous radioactive materials. The United States has traveled nearly 25 years down the current path only to come to a point where continuing to rely on the same approach seems destined to bring further controversy, litigation, and protracted delay.

The Blue Ribbon Commission on America's Nuclear Future (the Commission) was chartered to recommend a new strategy for managing the back end of the nuclear fuel cycle. We approached this task from different perspectives but with a shared sense of urgency. Put simply, this nation's failure to come to grips with the nuclear waste issue has already proved damaging and costly and it will be more damaging and more costly the longer it continues: damaging to prospects for maintaining a potentially important energy supply option for the future, damaging to state-federal relations and public confidence in the federal government's competence, and damaging to America's standing in the world—not only as a source of nuclear technology and policy expertise but as a leader on global issues of nuclear safety, non-proliferation, and security. Continued stalemate is also costly—to utility ratepayers, to communities that have become unwilling hosts of long-term nuclear waste storage facilities, and to U.S. taxpayers who face mounting liabilities, already running into billions of dollars, as a result of the failure by both the executive and legislative branches to meet federal waste management commitments.

The need for a new strategy is urgent, not just to address these damages and costs but because this generation has a fundamental ethical obligation to avoid burdening future generations with the entire task of finding a safe permanent solution for managing hazardous nuclear materials they had no part in creating. At the same time, we owe it to future generations to avoid foreclosing options wherever possible so that they can make choices—about the use of nuclear energy as a low-carbon energy resource and about the management of the nuclear fuel cycle—based on emerging technologies and developments and their own best interests.

Almost exactly one year after the Commission was chartered and less than five months before our initial draft report was due, an unforeseen event added yet more urgency to our charge and brought the problem of nuclear waste into the public eye as never before. A massive earthquake off the northeastern coast of Japan and the devastating tsunami that followed set off a chain of problems at the Fukushima Daiichi nuclear power station that eventually led to the worst nuclear accident since Chernobyl. In the weeks of intense media coverage that followed, many Americans became newly aware of the presence of tens of thousands of tons of spent fuel at more than 70 nuclear power plant sites around this country—and of the fact that the United States currently has no physical capacity to do anything with this spent fuel other than to continue to leave it at the sites where it was first generated. . . .

A NEW CONSENT-BASED APPROACH TO SITING

Siting storage or disposal facilities has been the most consistent and most intractable challenge for the U.S. nuclear waste management program. Of course, the first requirement in siting any facility centers on the ability to demonstrate adequate protection of public health and safety and the environment. Beyond this threshold criterion, finding sites where all affected units of government, including the host state or tribe, regional and local authorities, and the host community, are willing to support or at least accept a facility has proved exceptionally difficult. The erosion of trust in the federal government's nuclear waste management program has only made this challenge more difficult. And whenever one or more units of government are opposed, the odds of success drop greatly. The crux of the challenge derives from a federal/state/tribal/local rights dilemma that is far from unique to the nuclear waste issue—no simple formula exists for resolving it. Experience in the United States and in other nations suggests that any attempt to force a top-down, federally mandated solution over the objections of a state or community—far from being more efficient—will take longer, cost more, and have lower odds of ultimate success.

By contrast, the approach we recommend is explicitly adaptive, staged, and consent-based. Based on a review of successful siting processes in the United States and abroad—including most notably the siting of a disposal facility for transuranic radioactive waste, the Waste Isolation Pilot Plant (WIPP) in New Mexico, and recent positive outcomes in Finland, France, Spain and Sweden—we believe this type of approach can provide the flexibility and sustain the public trust and confidence needed to see controversial facilities through to completion.

In practical terms, this means encouraging communities to volunteer to be considered to host a new nuclear waste management facility while also allowing for the waste management organization to approach communities that it believes can meet the siting requirements. Siting processes for waste management facilities should include a flexible and substantial incentive program.

The approach we recommend also recognizes that successful siting decisions are most likely to result from a complex and perhaps extended set of negotiations between the implementing organization and potentially affected state, tribal, and local governments, and other entities. It would be desirable for these negotiations to result in a partnership agreement or some other form of legally enforceable agreement with the organization to ensure that commitments to and by host states, tribes, and communities are upheld. All affected levels of government must have, at a minimum, a meaningful consultative role in important decisions; additionally, both host states and tribes should retain—or where appropriate, be delegated—direct authority over aspects of regulation, permitting, and operations where oversight below the federal level can be exercised effectively and in a way that is helpful in protecting the interests and gaining the confidence of affected communities and citizens. At the same time, host state, tribal and local governments have responsibilities to work productively with the federal government to help advance the national interest. . . .

TYING IT TOGETHER?

The overall record of the U.S. nuclear waste program has been one of broken promises and unmet commitments. And yet the Commission finds reasons for confidence that we can turn this record around. To be sure, decades of failed efforts to develop a repository for spent fuel and high-level waste have produced frustration and a deep erosion of trust in the federal government. But they have also produced important insights, a clearer understanding of the technical and social issues to be resolved, and at least one significant success story—the WIPP facility in New Mexico. Moreover, many people have looked at aspects of this record and come to similar conclusions.

The problem of nuclear waste may be unique in the sense that there is wide agreement about the outlines of the solution. Simply put, we know what we have to do, we know we have to do it, and we even know how to do it. Experience in the United States and abroad has shown that suitable sites for deep geologic repositories for nuclear waste can be identified and developed. The knowledge and experience we need are in hand and the necessary funds have been and are being collected. Rather the core difficulty remains what it has always been: finding a way to site these inherently controversial facilities and to conduct the waste management program in a manner that allows all stakeholders, but most especially host states, tribes and communities, to conclude that their interests have been adequately protected and their well-being enhanced—not merely sacrificed or overridden by the interests of the country as a whole.

This is by no means a small difficulty—in fact, many other countries have not resolved this problem either. However, we have seen other countries make significant progress with a flexible approach to siting that puts a high degree of emphasis on transparency, accountability, and meaningful consultation. We have had more than a decade of successful operation of WIPP. And most recently, we have witnessed an accident that has reminded Americans that we have little physical capacity at present to do anything with spent nuclear fuel other than to leave it where it is. Against this backdrop, the conditions for progress are arguably more promising than they have been in some time. But we will only know if we start, which is what we urge the Administration and Congress to do, without further delay.

NUCLEAR ENERGY INSTITUTE

"NUCLEAR ENERGY: POWERING AMERICA'S FUTURE," 2013

In 2012 the Nuclear Regulatory Commission (NRC) issued construction licenses for new commercial nuclear reactors in South Carolina and Georgia—the first such approvals issued since before the 1979 accident at Three Mile Island. Although the plants have faced some opposition, construction is proceeding. These developments seem to promise a rebirth of commercial nuclear power. In this brochure the industry policy organization Nuclear Energy Institute makes the case for nuclear power for environmental and economic reasons. Why does it tout these priorities? In what ways do the claims advanced here about nature, science, and progress parallel or diverge from similar claims in the past?

AMERICA'S LARGEST SOURCE OF EMISSION-FREE ELECTRICITY

Nuclear energy is the largest and most efficient source of carbon-free electricity and the only one that can be expanded widely. On average, reactors operate about 90 percent of the time, powering America's homes and businesses 24/7, 365 days a year.

Excerpted from Nuclear Energy Institute, "Nuclear Energy: Powering America's Future," November 2013, www.emagcloud.com/et/Powering_Americas_Future_Version_2013/index.html.

Carbon Dioxide Emissions by U.S. Electric Industry (2011)
(All fuels in million metric tons of CO_2)

	PERCENTAGE OF CO_2-FREE ELECTRICITY	AVERAGE CAPACITY FACTOR
Nuclear Energy	63.9%	86%
Hydropower	22.6%	48.3%
Geothermal	1.4%	69.2%
Wind Power	11.7%	31.3%
Solar Energy	0.4%	27%

SAFE, CLEAN AND RELIABLE

Nuclear energy produces electricity reliably and without emitting air pollution or greenhouse gases. Without it, levels of harmful emissions in the air would be much higher—particularly those that cause acid rain and smog.

A nuclear energy facility's life-cycle carbon footprint—including everything from mining uranium to operations to dismantling a retired plant—is comparable to wind and geothermal power and lower than solar, hydro and natural gas power plants.

Across the industry, nuclear energy facilities produce electricity around the clock, more efficiently than any other source of electricity. . . .

BUILDING STATE-OF-THE-ART REACTORS

Five large reactors are under construction in the Southeast. Georgia Power Co. is building two reactors at its Vogtle nuclear facility in Georgia, and SCANA is building two at its V.C. Summer site in South Carolina. The Tennessee Valley Authority is completing the Watts Bar 2 reactor in Tennessee.

Moreover, the Vogtle 3 and 4 reactors provide at least $4 billion more value to customers than the next best available technology, according to the Georgia Public Service Commission staff.

A single large reactor can generate enough electricity to supply a city the size of Atlanta or Boston.

The nuclear energy industry is developing innovative small reactors that can be built in modules, providing flexibility in adding electricity to the grid as it is needed. Small reactors also could provide stand-alone power to desalinate or purify water or to create process heat for industrial applications. These innovative designs are expected to enter the market around 2020.

WHY DOES AMERICA NEED NEW REACTORS?

Several factors have increased interest in nuclear energy, including environmental protection, the historical price volatility of natural gas and electricity demand growth. Maintaining a diverse supply of fuels helps balance the benefits, risks and costs associated with producing electricity.

EXPANDING THE ELECTRICITY SYSTEM

By 2040, America will need hundreds of new power plants to meet increasing electricity demand. The electric power sector also faces a $1 trillion investment in updating an aging transmission system, installing new environmental controls, and continuing to invest in efficiency.

State governments and the federal government have key roles to play in creating the conditions under which the electric power industry can make these investments. Georgia and South Carolina, for example, have policies that allow companies in those states to recover the cost of building new reactors as construction proceeds. *Because this pay-as-you-go method reduces financing cost, consumers will pay less for electricity from the nuclear energy facilities.*

BOOSTING STATE AND LOCAL ECONOMIES

The U.S. nuclear energy industry supports more than 100,000 quality, high-paying American jobs. The development of new reactors already has spurred the hiring of thousands of workers in engineering, manufacturing, services and other areas. The industry will need more workers to build these reactors and up to 700 permanent staff at each site to operate them.

And because of expected retirements, the industry will need to hire as many as 20,000 highly skilled workers by 2018 to operate and maintain existing reactors.

The nuclear energy industry has a rich tradition of hiring military veterans and has intensified its recruiting efforts to expand opportunities for women and minorities. The industry's goal is to be as diverse as the communities it serves.

ENERGY FOR THE FUTURE

Electricity is integral to modern life. It provides energy for home lighting, appliances and our interconnected digital communications. It powers factories, hospitals, office equipment, elevators and more. America's electricity supply system is so reliable that we sometimes take it for granted.

The U.S. Department of Energy says America will need 28 percent more electricity by 2040—and that means building hundreds of new facilities to generate electric power.

Americans need reliable sources of electricity and they want clean air. With nuclear energy, they can have both. In fact, nuclear energy has a set of attributes no other electricity source can match:

- High efficiency and reliability
- Low production cost
- No greenhouse gas emissions
- Large amount of electricity production 24/7. . . .

SAFELY MANAGING USED NUCLEAR FUEL

Used nuclear fuel consists of ceramic pellets inside metal tubes that are bundled to form fuel assemblies. Nuclear energy facilities transfer used fuel to steel-lined concrete vaults filled with water. After cooling for several years, it can be stored in massive steel or steel-lined concrete containers at the plant site.

In 2012, a blue ribbon commission appointed by President Obama recommended developing consolidated storage facilities for used nuclear fuel in communities that consent to host them. The commission also recommended creating a new, independent organization to manage the nation's used nuclear fuel program and to foster efforts to develop one or more permanent disposal facilities.

All of the used nuclear fuel produced over the past 50 years would cover a single football field as high as the goal posts.

KEN CALDEIRA, KERRY EMANUEL,
JAMES HANSEN, AND TOM WIGLEY

"TO THOSE INFLUENCING ENVIRONMENTAL POLICY BUT OPPOSED TO NUCLEAR POWER," 2013

In 2013 four prominent climate scientists—including James Hansen, the former NASA scientist who had emerged as the most prominent American scientific voice in the climate change debate—published this open letter to the environmental community, pleading for support for nuclear power. How do the authors come to this conclusion? In what ways do they respond to past concerns about the safety and reliability of nuclear power? How does their support for nuclear power compare to, and differ from, other voices from the Cold War era?

As climate and energy scientists concerned with global climate change, we are writing to urge you to advocate the development and deployment of safer nuclear energy systems. We appreciate your organization's concern about global warming, and your advocacy of renewable energy. But continued opposition to nuclear power threatens humanity's ability to avoid dangerous climate change.

We call on your organization to support the development and deployment of safer nuclear power systems as a practical means of

Letter from Ken Caldeira, Kerry Emanuel, James Hansen, and Tom Wigley, "To Those Influencing Environmental Policy but Opposed to Nuclear Power," 2013, http://dotearth.blogs.nytimes.com/2013/11/03/to-those-influencing-environmental-policy-but-opposed-to-nuclear-power/?_r=0.

addressing the climate change problem. Global demand for energy is growing rapidly and must continue to grow to provide the needs of developing economies. At the same time, the need to sharply reduce greenhouse gas emissions is becoming ever clearer. We can only increase energy supply while simultaneously reducing greenhouse gas emissions if new power plants turn away from using the atmosphere as a waste dump.

Renewables like wind and solar and biomass will certainly play roles in a future energy economy, but those energy sources cannot scale up fast enough to deliver cheap and reliable power at the scale the global economy requires. While it may be theoretically possible to stabilize the climate without nuclear power, in the real world there is no credible path to climate stabilization that does not include a substantial role for nuclear power[.]

We understand that today's nuclear plants are far from perfect. Fortunately, passive safety systems and other advances can make new plants much safer. And modern nuclear technology can reduce proliferation risks and solve the waste disposal problem by burning current waste and using fuel more efficiently. Innovation and economies of scale can make new power plants even cheaper than existing plants. Regardless of these advantages, nuclear needs to be encouraged based on its societal benefits.

Quantitative analyses show that the risks associated with the expanded use of nuclear energy are orders of magnitude smaller than the risks associated with fossil fuels. No energy system is without downsides. We ask only that energy system decisions be based on facts, and not on emotions and biases that do not apply to 21st century nuclear technology.

While there will be no single technological silver bullet, the time has come for those who take the threat of global warming seriously to embrace the development and deployment of safer nuclear power systems as one among several technologies that will be essential to any credible effort to develop an energy system that does not rely on using the atmosphere as a waste dump.

With the planet warming and carbon dioxide emissions rising faster than ever, we cannot afford to turn away from any technology that has the potential to displace a large fraction of our carbon emissions. Much has changed since the 1970s. The time has come for a fresh approach to nuclear power in the 21st century.

We ask you and your organization to demonstrate its real concern about risks from climate damage by calling for the development and deployment of advanced nuclear energy.

LATUFF CARTOONS

FUKUSHIMA CARTOON, 2014

The 2011 disaster at Japan's Fukushima Daiichi power plant brought back many old concerns about the safety and wisdom of commercial nuclear power. Political cartoonists generated hundreds of images connecting the disaster to the bombing of Japan during World War II, the accident at Three Mile Island, fear of radiation exposure, and other elements of the nuclear past. Did images like this one, published by freelance cartoonist Carlos Latuff, change peoples' minds, or reaffirm previously held convictions?

Courtesy of Latuff Cartoons, 2014.

FUKUSHIMA
UNITED
STATES
LATUFF 2014-OPERAMUNDI

JOHN ASAFU-ADJAYE ET AL.

"AN ECOMODERNIST MANIFESTO," 2015

Since the early 2000s, the term "Anthropocene" has come into wide use to describe a new geologic epoch, one in which human activities serve as a driving force for global ecological change. The authors of "An Ecomodernist Manifesto"—who work with an American think tank that tries to shape national and international conversations about energy, technology, and nature—use the concept of the Anthropocene as a starting point to assess some of the basic assumptions of American environmental thought. In this excerpt, they suggest that only technological innovation and intensification can foster both human development and a healthy environment, without compromising either of these goals. In making this claim, the authors tie together relationships with nature, the shape of civil society, and ideas about progress. What role do nuclear and other technologies play in their vision? How does this vision compare to past support for nuclear technologies—and earlier ideas about nature, civil society, and progress?

. . . .

To say that the Earth is a human planet becomes truer every day. Humans are made from the Earth, and the Earth is remade by human hands. Many earth scientists express this by stating that the Earth has entered a new geological epoch: the Anthropocene, the Age of Humans.

Excerpted from John Asafu-Adjaye et al., "An Ecomodernist Manifesto," (full text available at www.ecomodernism.org), pp. 6–9, 11, 20–24, 28, 30–31. Courtesy of Michael Shellenberger.

As scholars, scientists, campaigners, and citizens, we write with the conviction that knowledge and technology, applied with wisdom, might allow for a good, or even great, Anthropocene. A good Anthropocene demands that humans use their growing social, economic, and technological powers to make life better for people, stabilize the climate, and protect the natural world.

In this, we affirm one long-standing environmental ideal, that humanity must shrink its impacts on the environment to make more room for nature, while we reject another, that human societies must harmonize with nature to avoid economic and ecological collapse.

These two ideals can no longer be reconciled. Natural systems will not, as a general rule, be protected or enhanced by the expansion of humankind's dependence upon them for sustenance and well-being.

Intensifying many human activities—particularly farming, energy extraction, forestry, and settlement—so that they use less land and interfere less with the natural world is the key to decoupling human development from environmental impacts. These socioeconomic and technological processes are central to economic modernization and environmental protection. Together they allow people to mitigate climate change, to spare nature, and to alleviate global poverty. . . .

Personal, economic, and political liberties have spread worldwide and are today largely accepted as universal values. Modernization liberates women from traditional gender roles, increasing their control of their fertility. Historically large numbers of humans—both in percentage and in absolute terms—are free from insecurity, penury, and servitude.

At the same time, human flourishing has taken a serious toll on natural, nonhuman environments and wildlife. . . .

Given that humans are completely dependent on the living biosphere, how is it possible that people are doing so much damage to natural systems without doing more harm to themselves?

The role that technology plays in reducing humanity's dependence on nature explains this paradox. Human technologies, from those that first enabled agriculture to replace hunting and gathering, to those that drive today's globalized economy, have made humans less reliant upon the many ecosystems that once provided their only sustenance, even as those same ecosystems have often been left deeply damaged. . . .

Even as human environmental impacts continue to grow in the aggregate, a range of long-term trends are today driving significant decoupling of human well-being from environmental impacts.

Decoupling occurs in both relative and absolute terms. *Relative* decoupling means that human environmental impacts rise at a slower rate than overall economic growth. Thus, for each unit of economic output, less environmental impact (e.g., deforestation, defaunation, pollution) results. Overall impacts may still increase, just at a slower rate than would otherwise be the case. *Absolute* decoupling occurs when total environmental impacts—impacts in the aggregate—peak and begin to decline, even as the economy continues to grow. . . .

Plentiful access to modern energy is an essential prerequisite for human development and for decoupling development from nature. The availability of inexpensive energy allows poor people around the world to stop using forests for fuel. It allows humans to grow more food on less land, thanks to energy-heavy inputs such as fertilizer and tractors. Energy allows humans to recycle waste water and desalinate sea water to spare rivers and aquifers. It allows humans to cheaply recycle metal and plastic rather than to mine and refine these minerals. Looking forward, modern energy may allow the capture of carbon from the atmosphere to reduce the accumulated carbon that drives global warming. . . .

There remains much confusion, however, as to how this might be accomplished. In developing countries, rising energy consumption is tightly correlated with rising incomes and improving living standards. Although the use of many other material resource inputs such as nitrogen, timber, and land are beginning to peak, the centrality of energy in human development and its many uses as a substitute for material and human resources suggest that energy consumption will continue to rise through much if not all of the 21st century.

For that reason, any conflict between climate mitigation and the continuing development process through which billions of people around the world are achieving modern living standards will continue to be resolved resoundingly in favor of the latter.

Climate change and other global ecological challenges are not the most important immediate concerns for the majority of the world's people. Nor should they be. A new coal-fired power station in Bangladesh may bring air pollution and rising carbon dioxide emissions but will also save lives. For millions living without light and forced to burn dung to cook their food, electricity and modern fuels, no matter the source, offer a pathway to a better life, even as they also bring new environmental challenges.

Meaningful climate mitigation is fundamentally a technological challenge. By this we mean that even dramatic limits to per capita global consumption would be insufficient to achieve significant climate mitigation. Absent profound technological change there is no credible path to meaningful climate mitigation. While advocates differ in the particular mix of technologies they favor, we are aware of no quantified climate mitigation scenario in which technological change is not responsible for the vast majority of emissions cuts. . . .

. . . Transitioning to a world powered by zero-carbon energy sources will require energy technologies that are power dense and capable of scaling to many tens of terawatts to power a growing human economy.

Most forms of renewable energy are, unfortunately, incapable of doing so, the scale of land use and other environmental impacts necessary to power the world on biofuels or many other renewables are such that we doubt they provide a sound pathway to a zero-carbon low-footprint future. . . .

Nuclear fission today represents the only present-day zero-carbon technology with the demonstrated ability to meet most, if not all, of the energy demands of a modern economy. However, a variety of social, economic, and institutional challenges make deployment of present-day nuclear technologies at scales necessary to achieve significant climate mitigation unlikely. A new generation of nuclear technologies that are safer and cheaper will likely be necessary for nuclear energy to meet its full potential as a critical climate mitigation technology.

In the long run, next-generation solar, advanced nuclear fission, and nuclear fusion represent the most plausible pathways toward the joint goals of climate stabilization and radical decoupling of humans from nature. . . .

The ethical and pragmatic path toward a just and sustainable global energy economy requires that human beings transition as rapidly as possible to energy sources that are cheap, clean, dense, and abundant. Such a path will require sustained public support for the development and deployment of clean energy technologies, both within nations and between them, th[r]ough international collaboration and competition, and within a broader framework for global modernization and development. . . .

Too often, modernization is conflated, both by its defenders and critics, with capitalism, corporate power, and laissez-faire economic policies. We reject such reductions. What we refer to when we speak of modernization is the long-term evolution of social, economic, political, and technological arrangements in human societies toward vastly improved material well-being, public health, resource productivity, economic integration, shared infrastructure, and personal freedom. . . .

Accelerated technological progress will require the active, assertive, and aggressive participation of private sector entrepreneurs, markets, civil society, and the state. While we reject the planning fallacy of the 1950s, we continue to embrace a strong public role in addressing environmental problems and accelerating technological innovation, including research to develop better technologies, subsidies, and other measures to help bring them to market, and regulations to mitigate environmental hazards. And international collaboration on technological innovation and technology transfer is essential in the areas of agriculture and energy. . . .

We offer this statement in the belief that both human prosperity and an ecologically vibrant planet are not only possible but also inseparable. By committing to the real processes, already underway, that have begun to decouple human well-being from environmental destruction, we believe that such a future might be achieved. As such, we embrace an optimistic view toward human capacities and the future.

INDEX

A

Abalone Alliance, 204–6
Abraham, Spencer, 279
acid rain, 12, 249, 297
AEC (Atomic Energy Commission), 10, 81, 96, 111, 149, 155, 163, 165, 172–73, 175; and commercial nuclear power, 14, 83, 97–99, 149–51; and environmental issues, 10, 47, 79, 147, 179–80; and industrial research, 82, 118–22; and nuclear testing, 111–13; and thermonuclear weapons, 49, 54, 56
Afghanistan, 263, 274–75
Africa, 189, 256, 274
Alamogordo, NM, xiii, 10
Alamogordo Air Base, 23
Alaska, 251
Albright, David, 273–75
Albuquerque, 24
All-Atomic Comics, 196–97
Al Qaeda, 273–75
alternative energy. *See* renewable energy
American University, 152
American West, 111
americium-241, 4
Ames Research Center, NASA, 241
Anthropocene, xiii, xvi, 305–6
Anthropocene Working Group, xiii
anticommunism, 7, 65, 86, 104, 106, 221
antinuclear movement, 9, 12–15, 163, 194–98, 200–206, 214, 222, 230, 251–54, 271; and environmental concerns, 12–13, 123, 171–79, 230–33
Arabian Nights, 88
arms control, 13, 17, 78, 116, 165, 178, 185, 218, 222, 224, 234–35, 237, 251, 263, 265. *See also* disarmament
arms race, 8, 13–14, 17, 54, 57, 63, 65, 75, 101, 132, 143, 154, 160–61, 177, 200–201, 222, 231–32, 268–69; environmental and social costs, 13, 101, 103; protest against, 13, 103–6, 114–17, 144, 218, 230, 238, 242, 251–54
Asafu-Adjaye, John, 305–9
Asia, 61, 189
Atlanta, GA, 297
Atomic Age: dawn of, xiii-xiv, 19, 21, 26, 32, 49, 149, 235, 271; new beginning, 193; and new technology, 84, 87, 95; and nuclear thought, 75, 103, 200–201. *See also* Nuclear Age
atomic bombs: and the arms race, 36, 44, 49, 59, 75; and civil defense, 65–67; and civil society, 37–38; and commercial nuclear power, 78, 220; development of, xiii, 23, 28–29, 41–42, 49; and the nuclear paradox, 4, 28–29; and popular culture, 32–33, 39; and progress, 35, 44, 157; and public anxiety, 201; and thermonuclear weapons, 50–53, 55–56; and World War II, 4, 29. *See also* atomic weapons; nuclear weapons
atomic energy, 8–9, 30, 34–38, 43, 110, 156, 173, 218; development of, 29, 84; and government, 34, 82, 100; peaceful uses, 8, 74–78, 82–88, 90, 92, 97–98; and weapons, 43, 82, 227. *See also* nuclear energy

Atomic Energy Act, 34, 81, 83, 193
Atomic Energy Commission. *See* AEC
Atomic Industrial Forum, 119, 183
Atomic Man, 39–40
atomic power, 30–31, 35–36, 41–42, 75, 98, 217. *See also* nuclear energy; atomic energy
atomic weapons, 7, 28, 61–63, 111–12, 145, 201. *See also* nuclear weapons
Atoms for Peace, 14, 74–78, 81, 96–100; and civil society, 9; and commercial nuclear power, 97–100; and consumer society, 8–9, 118–22; and the nuclear consensus, 8–9, 96; and progress, 9
Auschwitz, 222

B

baby tooth survey, 11, 144
Bainbridge, Kenneth, 25
Bangladesh, 307
Berlin, 265
Berlin Crisis, 152
Berlin Wall, 17
B-52 bomber, 268
Bhagavad Gita, xvi
Bikini Atoll, 101, 104, 240
Bin Laden, Osama, 273–74
biology, 108, 113, 124–25, 140, 240
Birks, John, 241
bomb shelter. *See* fallout shelter
Born, Max, 106
Boulder City, NV, 260
Bravo Test. *See* Castle Bravo Test
"breeder" reactors, 192, 194
Brezhnev, Leonid, 223
Bridgman, Percy W., 106
British Empire, 59
British Parliament, 221
Brower, David, 12
B-29 bomber, 66
Buckley, Oliver E., 52
Buehler, Kathryn, 273–75
Bulletin of the Atomic Scientists, 267–70
Bunkerville, NV, 212
Bureau of Public Roads, 93–95
Bush, George, 268
Bush, Vannevar, 27

C

Caldeira, Ken, 300–302
Caldicott, Helen, 200–203
California, 15, 204–6, 260, 289
Calvert Cliffs' Coordinating Committee, 179–82
Cambodia, 222
Campaign for a Nuclear Free Future, 251–54
Canada, 43
cancer: as a consequence of radioactivity, 125, 147, 162, 172–74, 176, 201–2, 205, 212–13, 252; radioactivity as a treatment for, 9, 82
carbon dioxide, 84, 249, 296–97, 301, 307.
carbon-14, 84, 120, 136, 176
carbon monoxide, 243
Carlsbad, NM, 260
Carson, Rachel, 11–12
Carter, Jimmy, 207
Castle Bravo Test, 101, 103, 240
Central America, 263
Central Park, 16
cesium, 176
Chazov, Yevgeny, 255
chemistry, 10, 108, 131
Chernobyl, 293
China, 240
Chinook salmon, 283
CIA, 202, 273
citizenship, xvi, 6, 9, 18, 57, 65, 101, 140, 196–97, 203
civil defense, 8–9, 65–73, 93–95, 112, 132, 135; and consumer society, 8–9,

140, 93–95; environmental considerations, 138; and fallout shelters, 140–43, 201; and gender roles, 8–9, 65, 69–73
civil rights movement, 14, 201
civil society, xv, xvii, 5–9, 15, 17–19, 21, 34, 44, 60–61, 69, 101, 124–26, 196, 203, 215, 221, 251, 272, 293–95, 305, 309; and centralized government, xv, 9, 12, 15, 36–37, 41, 205; and the consumer economy, 6, 8; and civil defense, 65, 140; and the nuclear consensus, 15, 101
Clamshell Alliance, 198
climate change, xvi, 3–5, 12, 17–19, 249, 271–72, 277, 279, 288, 300–302, 306–7
climate mitigation, 19, 307–308
climatic effects of nuclear war, 241, 244, 249, 256
coal, 30, 42, 83, 91, 184, 193, 247, 249, 307
cobalt, 82
Cohen, Lizabeth, 6
Cold War, 5–9, 13, 54, 57, 66, 74, 86, 103, 107, 140, 221, 230, 263; end of, 16–17, 19, 215, 263, 267–68, 282; and the nuclear consensus, 8, 57; nuclear tensions, 11, 18, 59, 63–64, 97, 134, 140, 152–53, 160, 215, 255
Colorado River, 176
Columbia River, 45–47, 172, 283, 285
commercial nuclear power. *See* nuclear power
Committee for the Present Danger (CPD), 188–91
Commoner, Barry, 11, 20, 123–27, 128, 144, 194
communism, 6, 8–9, 57, 104, 153, 160, 271
Communist Party, 13, 36
Compton, Arthur H., 41
Conant, James B., 27, 52
Consolidated Edison, 83
consumer economy, 9, 14
consumer goods, 6–7, 9, 57, 118, 137
consumerism, 6, 8
consumer society, xv, 6–9, 12, 14, 115, 118, 219–20; environmental costs of, 6–7, 12, 18, 175
Consumers' Republic, 6, 8
containment, 6–7, 57, 59, 62
CPD. *See* Committee for the Present Danger
Crutzen, Paul, 241
Cuban Missile Crisis, 13–14, 16, 101, 152, 155, 160
Curie, Marie, 21

D

Dachau, 222
Dartmouth College, 207
DDT, 176
Death Valley, CA, 178
decoupling, 306–308
DeLucchi, Mark, 289
Denver, CO, 177
Department of Defense, 56, 239
Department of Energy, 259–60, 278–80, 282–85, 299
Department of State, 55–56
Department of Interior, 139
détente, 163, 188–89, 221
deterrence: and civil defense, 71, 141–43; and "mutually assured destruction," 57, 291; policy of, 4, 7, 17, 52, 57, 62–63, 95, 141–43, 189–90, 239; and SDI, 234–36, 265
de Tocqueville, Alexis, 266
Deuster, Ralph W., 192–95
development: economic, 8, 15, 18, 86, 97, 110, 169, 264, 301; human, 17, 305–7, 309; nuclear research and, 7, 9, 36–38, 82–83, 122, 149–51, 163, 168, 190, 237
Diablo Canyon, 15, 204–206
disarmament, 14, 56, 116, 145, 156, 178, 201, 214, 254. *See also* arms control

Disney, Walt, 86–87
Disneyland, 87
Dixon, Bill, 284–86
DOE. *See* Department of Energy
Doomsday Clock, 267–70
"downwind" communities, 111, 211
Duquesne Light Co., 83

E

ecology, 10, 138, 199, 245, 286; development of, xvi, 10–12, 107; impact of nuclear technology on, 123, 134, 138–39, 230, 256
Ecomodernist Manifesto, 305–9
economic growth, 6, 8, 18, 59, 83, 95–98, 153, 163–64, 219, 221–22, 306–8
economic implications of nuclear energy, 16, 65, 93, 134, 136, 138–39, 179, 215, 233, 261, 211, 252, 288, 296, 306–9; of the Cold War, 57, 63, 188–91, 264; of commercial nuclear power, 8, 14–15, 57, 74, 77, 150–51, 167–69, 181, 184–86, 192, 205–6, 246–47, 250, 252; and the nuclear consensus, 15, 68, 163; of radioactive waste, 46–47, 225, 261. *See also* consumer society
ecosystem ecology. *See* ecology
ecosystems, 10, 138–39, 306
Einstein, Albert, 13, 103–6, 257
Eisenhower, Dwight D., 8, 57, 74–79, 81–83, 85, 158
electricity: costs, 83; demand for, 14, 167, 170, 185, 246–47, 298; nuclear generated, 4, 7, 19n2, 97, 118, 150, 163, 165–70, 178, 193, 249, 296–99. *See also* energy
Elk River, MN, 172
El Paso, TX, 24
Emanuel, Kerry, 300–302
energy conservation, 183–84, 186–87, 204–6, 219, 254
energy consumption, 150, 307–8. *See also* electricity
energy crisis (1973), 14, 163, 183–84, 219
energy policy, 185, 196, 205–6, 216, 218–20, 247, 249, 291, 307
Energy Research and Development Administration (ERDA, 163, 195
England, 221
environmental impacts: of commercial nuclear power, 57, 171–74, 201–6, 246; of fallout, 7, 11, 95, 125, 139, 144, 147, 160, 213; of nuclear energy in general, 7, 18, 101, 168, 179, 211; of nuclear weapons, 10, 35, 103, 239; of waste, 107. *See also* ecology
environmental movement: nuclear energy and the development of, 7, 10–12, 165–66, 179; relation to nuclear power, xv-xvi, 3, 12, 14, 17, 123, 166, 169, 175, 271, 278; and rethinking nuclear energy, xvi, 3, 17, 271, 288
environmental protection, xv, 180–82, 298, 306
Environmental Protection Agency (EPA), 227, 280, 283–85
ERDA. *See* Energy Research and Development Administration
ethics of nuclear energy, 49, 53–54, 124, 140, 177, 213, 220, 225, 292, 308. *See also* moral implications
Europe, 17, 141, 189, 222, 240, 244, 256, 265, 274

F

fallout, 7, 10–13, 24, 94, 101, 113, 123–27, 160–61, 176, 201, 211; and development of ecology, 10, 107, 123, 127; environmental impact of, 7, 11, 95, 125, 139, 144, 147, 213; human health implications, 213–14; and milk, 11, 144–48; public concern, 11–13, 17, 124, 126
fallout maps, 132–33

fallout shelters, 8, 72, 132, 140–43, 259, 201, 243
Farrell, Thomas F., 23, 25
Federal-Aid Highway Act, 93
Federal Civil Defense Administration (FCDA), 65–73, 94
Federal Power Commission, 98
Federal Radiation Council, 173
Fermi, Enrico, 53
Finland, 294
foreign policy, American, 7–8, 145, 163, 188–90, 200, 221, 263, 266
Fortune Magazine, 79
fossil fuels, xiii, xvi, 12, 167, 178, 185, 249, 289, 301; limits on, 83, 149–50, 163; price of, 14, 150
France, 294
freedom: economic, 7–8, 15, 188, 215, 264, 306; personal, 60, 153, 222, 309; political, 15, 37–39, 63, 76, 97, 118, 158, 184, 189–90, 221–24; and progress, 15, 41–42; religious, 264; from war, 42, 44, 264
French Empire, 59
fuel cycle, 186–87, 192–95, 253, 292
Fukushima Daiichi, 3, 272, 293, 303–4
Fundamentals of Ecology, 174

G

GAC. *See* General Advisory Committee
gender roles, 7, 9, 65, 69, 306
General Advisory Committee, 49–54
Geneva, Switzerland, 161, 235
geographical engineering, 128–131
geologic disposal of radioactive waste, 226–28, 259–60, 277–81, 291, 294–95
Georgia, 3, 296–98
Georgia Public Service Commission, 297
Germany, 59
Gofman, John, 175, 177
Gorbachev, Mikhail, 265, 268–69
Gould, William R., 183–87
government, role in development of nuclear energy, 82–83, 96, 98–100, 149–51, 186, 198, 207, 261, 292, 294, 298
Great Britain, 29, 43
Greater St. Louis Citizens' Committee for Nuclear Information, 11, 123, 144
greenhouse gasses, xvi, 272, 297, 299, 301. *See also* carbon dioxide
Gregerson, Gloria, 211–14
Groves, Leslie, 23–27
gulag, 222
Gulf War, 269

H

Haber, Heinz, 86–92
Hanford Federal Facility Agreement and Consent Order. *See* Tri-Party Agreement
Hanford Reach, 283
Hanford Works, 45–48, 172, 282–87
Hansen, James, 300–302
Harrisburg, PA, 207
Harvard Medical School, 200
H-bomb. *See* hydrogen bomb
health: environmental, 17–18, 107, 164, 175, 200, 255, 272, 305; human, xv, 8, 12–13, 48, 91–92, 101, 107, 128, 168–70, 214, 217, 255, 272; public, 202, 226, 256, 293, 309
Higgins, Holly, 273–75
Hiroshima, 4, 21, 28, 37, 49, 104, 155, 175, 213, 239, 256, 273
Hollywood, 114
Hoover, Herbert, 85
Hoover Dam, 261
House Un-American Activities Committee, 13
Hubbard, Jack M., 25
Hughes, Langston, 257

human rights, 154, 189, 223
hydrocarbons, 120, 193
hydroelectricity, 289–90, 297
hydrogen bomb, 13, 49–57, 75, 103–105, 114. *See also* thermonuclear weapons

I

India, 145, 202
industrial radiation chemistry, 9
Industrial Revolution, xiii, 221
INF treaty. See Intermediate-Range Nuclear Forces Treaty
Institute for Science and International Security, 274
Intermediate-Range Nuclear Forces Treaty (INF Treaty), 17, 263
International Atomic Energy Agency, 77
International Commission on Stratigraphy (ICS), xiii
international control of nuclear energy, 42, 44, 51, 53, 74, 77–78, 116
International Physicians for the Prevention of Nuclear War (IPPNW), 255–58
Iran, 4
isotopes, 4, 8, 10–12, 82, 84, 118–20, 131, 169, 172

J

Jacobson, Mark Z., 288–90
Japan: and the Castle-Bravo Test, 101, 104; and Fukushima Daichii, 3, 272, 293, 303; and World War II, xiii, 4, 11, 28, 30, 200. *See also* Hiroshima; Nagasaki
Joint Committee on Atomic Energy, 96, 99, 124
Joliot-Curie, Frederic, 106

K

Kahn, Herman, 134–39
Kemeny, John G., 207
Kemeny Commission, 207–10
Kennedy, John F., 13, 140, 145, 152–54, 158, 160–62
Kentucky, 253
Khrushchev, Nikita, 161
King Solomon, 88–89
Kistiakowsky, George, 25–26
Kremlin, 56, 60–61, 63–64

L

Las Vegas, NV, 259–62
Latin America, 189
Latuff, Carlos, 303–4
leukemia, 147, 162, 172–73, 214
Libby, Willard F., 120
Life magazine, 4, 32–33
Lilienthal, David E., 155–59, 217–20
Los Alamos, NM, 282
Los Angeles, CA, 214
low-carbon energy, 3, 17–18, 271, 288–89, 292, 296, 308
Lown, Bernard, 255–59
Lucky Dragon, 101, 103

M

Macfarlane, Allison M., 277–81
Macias, Elizabeth, 259–62
Magic Kingdom, 86
Mainz, West Germany, 241
Manhattan Project, xiv, 9, 14, 23, 41, 49, 128, 165, 267
Mann, Adam, 39–40
Mars, 241
Marshall, Bob, 175
Marshall, Lenore, 175–78
Marshall Islands, 240
Marxism-Leninism, 224

Massachusetts, 83
Mattson, Roger, 208
Max Planck Institute, 241
May-Johnson Bill, 34, 36
McCarthy, Joseph, 101
McCormick, Anne O'Hare, 4
Mead, Margaret, 140–43
media, 3, 28, 32–33, 101, 126, 145, 166, 186–87, 194, 209–10, 223, 293
medicine, 4, 8, 77, 79, 113, 124, 240, 256, 258
Merrill, David N., 198–99
Michigan, 83
Mickey Mouse, 86
Middle East, 138, 183, 189, 263
military spending, 15, 152, 188–89, 221, 235, 253, 256
milk: radioactivity and, 11, 212; and Women Strike For Peace campaign, 144–48
Minneapolis, MN, 171
Minnesota, 15, 171, 174
Minnesota Environmental Control Citizens Association (MECCA), 171–74
"Missile Gap," 152
missiles, 17, 114–16, 224, 234–37, 253, 263, 265, 268
Mississippi River, 171–74
Mitchell, H. H., 134–39
modernity, 21
Monticello Plant, 171–74
Moore, W. Henson, 284–86
moral implications of nuclear energy, 6, 36–38, 41, 202; and fallout shelters, 140–43; and nuclear testing, 124–27; and science, 203, 256; for U.S. foreign policy, 64, 97, 115–17, 190, 219–20, 233; and weapons, 53, 55, 124, 136–37, 158, 233
Moscow, 104, 161, 189, 254, 265
mushroom cloud, 10, 24, 32–33, 88, 111–12
mutation, as a consequence of radioactivity, 9, 115, 142, 172–74
mutually assured destruction, 134, 233, 271. *See also* deterrence

N

Nader, Ralph, 194
Nagasaki, 21, 32–33, 213, 256
NASA (National Aeronautics and Space Administration), 241, 300
National Academy of Sciences, 56, 107, 109
National Committee for a Sane Nuclear Policy. *See* SANE
National Environmental Policy Act. *See* NEPA
National Interstate and Defense Highways Act, 93
National Park System, 227
National Security Council. *See* NSC
National Wild and Scenic River System, 227
National Wilderness Preservation System, 227
National Wildlife Refuge System, 227
natural rights, 114–15
nature, human relations to, xvi, 11–13, 30, 79, 25–26, 128, 156, 203, 230–33, 305–9
Nautilus, 84, 87
Nebraska, 83
Needles, CA, 260
NEPA (National Environmental Policy Act), 179–82, 195
Nevada: and nuclear testing, 11, 111–13, 177, 211–12; and radioactive waste, 259–62, 277, 291
Nevada Test Organization, 113
Nevada Test Site, 11, 111–13, 211, 260, 282
New England, 83
New Hampshire, 15, 198–99

New Look military strategy, 57
New Mexico, xiii, 23, 25, 259–60, 282, 294–95
New York (state), 83
New York City, 16, 104, 251, 273
New Yorker, 230
New York Times, 4
Niagara Falls, NY, 27
Nobel Prize, 21, 41, 165, 240, 255
Non-Proliferation Treaty, 269
North Atlantic Treaty Alliance (NATO), 224
Northern States Power Company, 172–73
North Korea, 4
NRC (Nuclear Regulatory Commission), 3, 163, 208–10, 247, 296; and radioactive waste, 194–95, 225, 277, 280–81; and regulation, 3, 199, 209
NSC (National Security Council), 7, 59–64
NSC-68, 7–8, 59–64, 74, 188
NTA (nitrilotriacetic acid), 176
Nuclear Age, 4–5, 191, 255, 271, 290. *See also* Atomic Age
nuclear arms race. *See* arms race
nuclear bomb. *See* nuclear weapons
nuclear consensus: challenges to, 13, 17, 101, 103, 114, 144, 155, 165, 217; and consumer society, 7–8, 12, 69, 96; emergence of, 6–7, 10, 21, 57, 59, 65, 188; environmental consequences of, 7, 165, 211, 282; and progress, 10, 12, 18, 114; renewed, 15, 188, 215, 221
nuclear energy: and civil society, 5, 10, 21, 34, 101, 251, 272; and climate change, xvi, 12, 17, 19, 272, 288–90, 300–302, 308; and consumer society, 6–7, 79–80; in daily life, 6, 18–19, 21, 41; economic/industrial uses, 3–4, 6, 12, 14–16, 21, 57, 74, 81–82, 118–22, 155, 163–64, 167–69, 178, 184–85, 221, 296–99; environmental impacts, 11, 57, 101, 107, 125, 128, 131, 149–50, 164, 168, 172, 206, 272, 296; history, xiv, xvi, 3–5, 17, 291; and the nuclear paradox, xiv-xv, 4–5, 12, 17, 271; opposition to, 9, 12–13, 193, 200, 205; and popular culture, 39–40, 79–80, 86–92; and progress, 5, 10, 18, 21, 41, 79–80, 175, 217, 267; public opinion, 3–5, 14, 39, 271; and relationships to nature, xiv-xvi, 5, 7, 10–12, 21, 79–80, 175; and waste, 192, 225, 277, 292; and weapons, 11, 15, 54, 57, 155. *See also* atomic energy
Nuclear Energy Institute, 296–99
Nuclear Freeze, 16, 234–35, 251–54
Nuclear Free Zones, 251–52, 254
Nuclear Fuel Services, 192
nuclear paradox, xiv-xv, 4–5, 10, 12, 14, 17–19, 164, 271
nuclear physics, 87, 124, 131
nuclear power: and climate change, xvi, 3, 271, 288, 296, 300–302, 308; consumer economy and, xv, 7–8, 96, 98, 151, 246; and the energy crisis, 14, 183; and environmental concerns, xvi, 13, 165, 169, 171–74, 176, 179, 187, 205, 296, 300; and the federal government, 8–9, 14–16, 96, 100, 149–51, 183, 185–86, 207; growth of, 35, 83–84, 107, 149–50, 163, 184, 192; and ideas of progress, 218; limits on, 183–85, 192, 246, 248–49; and the nuclear paradox, xv, 14, 163; and nuclear weapons, 205, 253, 269, 289; protest against, 15, 171–74, 177–78, 187, 196–97, 200, 205–6, 252, 254; public perception of, 3, 15–17, 163, 170, 198, 208, 215, 246–50; regulation of, 16, 96, 149, 163, 183–84, 185–86, 198–99, 208–10, 215; and safety, 3, 168–69, 205, 207–10, 247, 293, 300–301, 303; and waste, xv, 107, 217, 214, 278, 293

nuclear reactors: "breeder" reactors, 192, 194; and commercial nuclear power, 5, 15, 83, 97–99, 163, 171, 178, 192, 200, 202, 246, 248–50, 296–98; development, 14, 99, 150, 247, 298; and proliferation, 202, 205; and protests, 198–99, 201–6; and regulation, 179, 198–99, 248; and research, 41, 82–84, 150, 119; and safety, 3–4, 169, 200–203, 207, 247–48, 257, 271; and waste, 18, 121, 129, 192, 225, 277
Nuclear Regulatory Commission. *See* NRC
Nuclear Test Ban Treaty. *See* Partial Nuclear Test Ban Treaty
nuclear testing, xiii, 13, 75, 77, 82, 111–13, 160–62, 202, 211–14, 243, 268; environmental implications, 10–11, 108, 160; and fallout, 10–11, 123–27, 139, 147–48, 176–77; protest of, 144–48, 206, 234, 252–53; and public concern, 11, 15, 116, 123, 145, 147–48; of thermonuclear weapons, 50, 53, 101, 103–4, 267
nuclear waste. *See* radioactive waste
Nuclear Waste Fund, 227
Nuclear Waste Policy Act (NWPA), 225–29, 259, 280, 292
Nuclear Waste Policy Amendments Act, 280
nuclear weapons, 153, 160, 172, 222; and arms control, 105–6, 178, 206, 215, 232, 263, 265, 268; and the arms race, 57, 82, 85, 200, 239; and civil society, 9, 140; and deterrence, 235, 268; development of, 23, 107, 282–83, 285–86, 289; economic implications, 85, 94; environmental consequences, 139, 182; and fallout, 176, 94, 139; and the nuclear consensus, 6, 94; and nuclear winter, 238–45; and the peace movement, 105–6, 116, 163, 234, 252–54; post-Cold War, 3–4, 17, 268–69, 273–75; and proliferation, 192, 202, 205, 289; and public concern, 4, 13–14, 118, 140, 282; and testing, 11, 13, 123, 127, 176–77, 206, 234, 243, 253, 267; and thermonuclear weapons, 49–56, 63, 104, 106, 128, 140, 240, 267; and waste, 253, 259. *See also* atomic weapons; thermonuclear weapons
nuclear winter, 16, 238–45, 256, 271

O

Oak Ridge National Laboratory, 79–80
Obama, Barack, 291, 299
Obama Administration, 291, 295
oceanography, 10, 107–9
Odum, Eugene P., 10, 174
Odum, Howard, 10
Office of Civil Defense and Mobilization, 132–33
Office of Technology Assessment, 246–50
oil, 30, 83, 91, 119–20, 131, 167, 219; and 1973 energy crisis, 14, 163, 183–84; environmental impacts of, 176, 249. *See also* petroleum
On Thermonuclear War, 134
Oppenheimer, J. Robert, 13, 20, 25–26, 51–52
Oregon, 172, 283–85
Oregon Department of Ecology, 284
Oregon Department of Energy, 282–87
Organization of Petroleum Exporting Countries, 14
OTA. *See* Office of Technology Assessment

P

Pacific Gas & Electric, 204
Pacific Ocean, 3, 10–11, 112, 131, 211
Parade magazine, 238

Parker, Herbert M., 45–48
Partial Nuclear Test Ban Treaty, 13, 15, 33, 163, 267
Pathfinder Plant, 172
peace movement, 13–14, 16, 144–48, 163, 215, 234
Pearl Harbor, 29, 37, 106
Pentagon, 24, 201
Persian Gulf, 219
petroleum, 119–20, 184, 193. *See also* oil
Philadelphia, PA, 266
physics, 10, 42, 108, 125, 156
Plowshare program, 128–31
plutonium, 41, 45, 165, 253, 283; and proliferation, 192, 253, 268, 274–75, 289; and reprocessing, 192–94; and waste, 45–47, 176–77, 283
Pollack, James B., 241
pollution, 6, 14, 109, 166–67, 193, 249, 307; air, 12, 55, 162, 288, 297, 307; radioactive, 7, 46, 176, 205; thermal, 205; water, 46–47
Pollution Control Agency, Minnesota, 173
popular culture, 21, 32–33, 39–40, 79–80, 86–92, 196–97, 303–4
postattack environment, 136–39, 271
post–Cold War world, xv, 4–5, 17, 268, 271, 273–75
Potsdam Conference, 30
Powell, Cecil F., 106
power plants. *See* nuclear reactors
Prairie Island, MN, 15, 171
President's Blue Ribbon Commission on America's Nuclear Future, 291–95
progress: changing definitions of, xiv-xvii, 5, 7, 10, 17–19, 106, 114, 118, 157–58, 169–70, 215, 255, 272, 296, 305; and civil society, 57; and consumer society, 6, 9, 12, 17–18, 57, 79–81, 98, 115, 164, 180, 215, 219–20, 230, 296; and nature, 17, 79–80, 106, 218, 230, 272; and science/technology, xv-xvi, 10, 15, 21, 35, 42–43, 47, 86–87, 118, 126, 128, 155, 167, 177, 230, 309
Project Plowshare, 128–31
proliferation, 160, 192, 202, 254, 268–69, 274–75, 289, 292, 301
protest, 144, 291; antinuclear, 14, 174, 198; and civil society, 174, 215, 218; direct action, 15, 198, 204–8. *See also* antinuclear movement
public perception: and commercial nuclear power, 3, 15–16, 163, 207–10, 215, 246–50; and nuclear fear, 14, 132, 142, 201; and waste, 227, 293
Public Service Company of New Hampshire, 198
Public Utilities Commission of New Hampshire, 199

R

Rabi, I. I., 53
Rabinowitch, Eugene, 14, 20
radiation: environmental impacts, xiii, xvi, 109, 125, 139, 174–75; and fallout, 125, 127, 139, 147–48; human health impacts, 21, 45, 91, 101, 173, 175–77, 199, 201, 205, 211–14, 240, 256, 279; industrial uses, 9, 21, 118, 121–22; and public anxiety, 3, 125, 207, 209–10, 303; and research, 21, 84, 125, 131, 139, 148, 169
Radiation Exposure Compensation Act, 211
radioactive waste, 15, 17, 45–48, 107–10, 195, 225–29, 259–62, 277–89, 291–95; and antinuclear protest, 200, 203, 252–53, 172–73, 291; and commercial nuclear power, 192–95, 206; environmental concern, 172–74, 176, 200, 205, 282–85; ocean disposal of, 107–10; shipment of, 8, 84, 168, 227, 254, 259–62, 283; unresolved dilemma of, xv, 15, 17, 217, 271–72, 288, 301

radiochemistry, 124
radiophosphorus, 84
radiovulcanization, 121
radium, 21, 82, 176
RAND Corporation, 134
reactors. *See* nuclear reactors
Reagan, Ronald, 15–16, 215, 221–24, 234–37, 263–66
Reagan Administration, 16, 238, 263
Red Cross, 72
regulation of nuclear power, 109, 180–81, 284, 286; broken regulatory environment, 16, 82, 163, 184–87, 194–95, 198–99, 215, 246–47, 288; federal responsibility for, 36, 96, 109, 149, 169, 183, 208–10, 215.
renewable energy, 19, 197, 206, 247, 288–89, 300, 308
Report of the President's Commission on the Accident at Three Mile Island (Kemeny Commission), 207–10
repository. *See* geologic disposal
reprocessing, 192–93, 195, 225
research, 8–10, 19, 43, 79, 82; and development of nuclear energy, 7, 122, 149, 190, 237; environmental, 10–11, 109–10, 112, 309; on fallout, 126–27, 145; on industrial applications, 99, 120–22, 130, 160; medical, 9, 82, 91, 169, 225; on nuclear weapons, 29–30, 49, 134, 138–39, 203, 252–54, 274, 282; scientific, 8–10, 84, 112, 203, 252
Reuther, Walter, 96–100
Revelle, Roger, 107–10, 128
Ribicoff, Abraham, 147
Rifas, Leonard, 196–97
Rockefeller Foundation, 34
Rocky Flats plant, 177, 282
Roentgen, Wilhelm Conrad, 21
Rongelap, 240
Roosevelt, Franklin Delano, 158
Rotblat, Joseph, 106
Rumsfeld, Donald, 278
Russell, Bertrand, 13, 103–6
Russell-Einstein Manifesto, 103–6
Russia, 36, 147, 266, 274–75; and the development of nuclear energy, 52, 55, 82; military power, 15, 66, 71, 115. *See also* Soviet Union.

S

safety: of commercial nuclear power, 16, 151, 168–69, 171–73, 157, 199–200, 203–4, 208–10, 246–48, 288, 292, 300–301, 303; of milk, 11, 144–48; and radiation, 45, 175; and radioactive waste, 47, 54, 226–27, 261, 277, 286, 293; and weapons, 70, 75, 113, 130, 143, 237, 285
Sagan, Carl, 238–45
Saint Paul, MN, 171
SANE, 114–17, 144, 175
San Luis Obispo, CA, 204
Santa Fe, NM, 24
SCANA, 297
Schaefer, Milner B., 107–10, 128
Schell, Jonathan, 230–33
science, xvi, 7, 10–12, 86–87, 107, 120, 123–27, 277, 288, 300; ethical questions, 37, 49, 177, 203; politicization of, 17, 123, 272; and progress, xv, 10, 21, 35, 37–38, 43–44, 86, 92, 128, 175, 221, 230, 257–58, 296; research, 8–10, 84, 107–10, 130–31, 137–39; and uncertainty, 123–27, 272; use of, 4, 123, 175, 177, 218, 238, 240; and weapons, 29–30, 103, 109, 203, 218, 237. *See also* technology
Scientific American, 289
scientific authority, 10, 15, 125–26, 177, 196, 200, 203, 211, 214, 255–56
SDI (Strategic Defense Initiative), 215, 234, 238, 265
Seaborg, Glenn T., 165–79, 171

Seabrook Station Nuclear Power Plant, 15, 198–99
Searing, Hudson R., 83
Seawolf, 84
Second World War, *see* World War II
security: national, 4, 64, 85, 93, 183, 194, 235–36, 273–74, 292; new meanings of, 114–15, 255, 267–69
Shevardnadze, Eduard, 264
Shippingport, PA, 83, 98
Shultz, George, 265
Sierra Club, 12
Silent Spring, 11–12,
Silver City, NM, 24
Simpson, John A., 268
Sioux Falls, SD, 172
Smith, Cyril Stanley, 52
Smith, Randy, 284–8
sociology, 124
solar energy, 178, 183, 187, 193, 205–6, 250, 289–90, 297, 301, 308
South Carolina, 296–98
Soviet Union, 6–8, 15, 37, 54, 86, 188–91, 244, 268; and arms control, 17, 145, 154, 215, 234, 236, 263, 265, 268; demise of, 3, 17, 265–66, 268, 273; development of nuclear energy, 7, 49, 51, 56, 59, 77, 114, 267, 286; military power, 14, 57, 59–64, 152, 235–37, 239–40. *See also* Russia
Spain, 294
spent fuel, 3, 7, 18, 121, 192–95, 225–28, 277–78, 291–95. *See also* radioactive waste
Spider-Man, 39
Stanford University, 288–89
Stanley, Roger, 284–86
"Star Wars." *See* SDI
Strategic Arms Limitation Treaty, 222, 268
Strategic Defense Initiative. *See* SDI
strategic weapons, 190. *See also* nuclear weapons
Strauss, Lewis L., 54–56, 81–85
Strontium-90, 11–12, 124, 144, 147, 176
super bomb. *See* H-bomb; thermonuclear weapons
Sweden, 294
Sykesville, MD, 251

T

Taliban, 274–75
Tamplin, Arthur, 175, 177
technology: and commercial nuclear power, 14, 57, 74, 81, 92, 118, 150–51, 165, 171, 218, 236, 246–50, 254, 309; and energy, 205, 208, 308; and the environment, xv-xvi, 7, 12, 18, 101, 163, 166, 169–70, 175, 177, 187, 201–3, 218, 301, 305–8; and government, 12, 34, 37–38, 99, 150; and the nuclear paradox, 4, 12; and progress, xiv-xv, 4, 9–10, 15, 21, 35, 43–45, 47, 86, 155, 168–69, 218, 220–21, 257, 306; public perception, 14, 28, 41, 248, 267, 292; and weapons, 56, 215, 218, 234, 236–37, 301. *See also* science
Teller, Edward, 128–31
Tennessee, 25, 253, 297
Tennessee Valley Authority, 297
terrorism, 4–5, 17, 19, 271, 273–75
Test Ban Treaty. *See* Partial Nuclear Test Ban Treaty
thermonuclear weapons, 49–56, 63, 128, 134, 140, 240, 244–45; and Project Plowshare, 128; and commercial energy, 131; and testing, 101, 240, 267
Three Mile Island (TMI), 3, 15, 207–10, 215, 217, 296, 303
Toon, Brian, 241
transportation, 8, 93–95, 129, 149, 151, 167–68, 252–53, 289; of radioactive waste, 227–28, 254, 259–62, 283
Trident submarines, 16
Trinity Test, xiii, 23–27

Tri-Party Agreement, 282–85
tritium, 4
Truman, Harry S., 7, 28–31, 32, 49, 54, 59, 82, 158
TTAPS, 241
Turco, Richard, 241, 244
20,000 Leagues Under the Sea, 86

U

Union Carbide and Carbon Corporation, 79–80
Union of Soviet Socialist Republics (U.S.S.R.). *See* Soviet Union
United Auto Workers, 96
United Nations, 8, 63, 74, 76–77, 83, 85, 116, 145, 161, 263
United Nations Universal Declaration of Human Rights, 223
United States: and the Cold War, 7–8, 65–68, 86; commercial nuclear power industry, 3, 149, 185, 246, 297–99; and consumer society, 6, 8, 18, 21, 119–22, 205, 219; and the development of nuclear energy, 3–5, 29–30, 43, 53, 75, 82, 86–87, 96–99; energy policy, 14, 184, 187, 193; and environmental concerns, xiv-xv, 7, 11, 179, 211; foreign policy, 3, 7–8, 17, 57, 59–64, 74–78, 86, 97, 145, 152, 154, 161, 183, 200, 215, 235, 237, 263–65, 268, 275; and military implications of nuclear energy, 3, 11, 30, 44, 51, 53–55, 62, 134, 137–38, 142, 152, 154–55, 158, 188–90, 236, 239–40, 244, 274; and nuclear testing, 11, 101, 111, 128, 147; and radioactive waste, 260, 278, 280–82, 291–95
United States Air Force, 67, 113
United States Army, 30, 45
United States Army Air Force, 28
United States Army Corps of Engineers, 139
United States Congress, 30–31, 77, 82–83, 193, 211, 234, 246, 253–54, 284; and environmental concern, 123–24, 126, 165, 179–82; and radioactive waste, 123–24, 126, 165, 179–82; and regulation, 30–31, 163, 169, 198
United States Constitution, 60, 158, 266
United States Geological Survey, 227
United States Navy, 87
University of California–Davis, 289
uranium, 41–42, 76, 289; mining, 172, 211, 252–53, 289, 297; and waste, 46, 176, 192; and weapons, 41, 268, 274–75
U.S. Mariner 9, 241
U.S.-Soviet relations, 13, 65, 74, 141, 152, 161, 163, 215, 223–24, 263–65
U.S.S.R. *See* Soviet Union.

V

Vandenberg, Hoyt, 67
V.C. Summer Nuclear Generating Station, 297,
Vietnam War, 14, 67, 163, 188, 201–2
Vogtle Electric Generating Plant, 297

W

War on Terror, 4–5, 19, 271
Washington (state), 25, 45, 282–85
Washington, DC, 161, 184, 235, 259, 264, 285–86
Washington Department of Ecology, 283–84
Waste Isolation Pilot Plant (WIPP), 259–60, 294–95
Watergate, 202
Watkins, James, 284, 286–87
Wellock, Thomas, 15
Wigley, Tom, 300–302
wilderness, 93, 175–78

Wilderness Society, 175
wildlife, 139, 306
Willits, Joseph H., 34–37
wind power, 216, 289–90, 297, 301
WIPP. *See* Waste Isolation Pilot Plant
Women Strike for Peace, 144–48
World Health Organization, 240
World War II, 21, 41, 128, 149, 153, 165, 239, 303
World War III, 71, 256
Worster, Donald, 10

X

X-Men, 39
X-rays, 4, 21, 120, 173, 176, 243

Y

Yucca Mountain, 259–61, 277–81, 291–92

WEYERHAEUSER ENVIRONMENTAL BOOKS

Defending Giants: The Redwood Wars and the Transformation of American Environmental Politics, by Darren Frederick Speece

The City Is More Than Human: An Animal History of Seattle, by Frederick L. Brown

Wilderburbs: Communities on Nature's Edge, by Lincoln Bramwell

How to Read the American West: A Field Guide, by William Wyckoff

Behind the Curve: Science and Politics of Global Warming, by Joshua P. Howe

Whales and Nations: Environmental Diplomacy on the High Seas, by Kurkpatrick Dorsey

Loving Nature, Fearing the State: Environmentalism and Antigovernment Politics before Reagan, by Brian Allen Drake

Pests in the City: Flies, Bedbugs, Cockroaches, and Rats, by Dawn Day Biehler

Tangled Roots: The Appalachian Trail and American Environmental Politics, by Sarah Mittlefehldt

Vacationland: Tourism and Environment in the Colorado High Country, by William Philpott

Car Country: An Environmental History, by Christopher W. Wells

Nature Next Door: Cities and Trees in the American Northeast, by Ellen Stroud

Pumpkin: The Curious History of an American Icon, by Cindy Ott

The Promise of Wilderness: American Environmental Politics since 1964, by James Morton Turner

The Republic of Nature: An Environmental History of the United States, by Mark Fiege

A Storied Wilderness: Rewilding the Apostle Islands, by James W. Feldman

Iceland Imagined: Nature, Culture, and Storytelling in the North Atlantic, by Karen Oslund

Quagmire: Nation-Building and Nature in the Mekong Delta, by David Biggs

Seeking Refuge: Birds and Landscapes of the Pacific Flyway, by Robert M. Wilson

Toxic Archipelago: A History of Industrial Disease in Japan, by Brett L. Walker

Dreaming of Sheep in Navajo Country, by Marsha L. Weisiger

Shaping the Shoreline: Fisheries and Tourism on the Monterey Coast, by Connie Y. Chiang

The Fishermen's Frontier: People and Salmon in Southeast Alaska, by David F. Arnold

Making Mountains: New York City and the Catskills, by David Stradling

Plowed Under: Agriculture and Environment in the Palouse, by Andrew P. Duffin

The Country in the City: The Greening of the San Francisco Bay Area, by Richard A. Walker

Native Seattle: Histories from the Crossing-Over Place, by Coll Thrush

Drawing Lines in the Forest: Creating Wilderness Areas in the Pacific Northwest, by Kevin R. Marsh

Public Power, Private Dams: The Hells Canyon High Dam Controversy, by Karl Boyd Brooks

Windshield Wilderness: Cars, Roads, and Nature in Washington's National Parks, by David Louter

On the Road Again: Montana's Changing Landscape, by William Wyckoff

Wilderness Forever: Howard Zahniser and the Path to the Wilderness Act, by Mark Harvey

The Lost Wolves of Japan, by Brett L. Walker

Landscapes of Conflict: The Oregon Story, 1940–2000, by William G. Robbins

Faith in Nature: Environmentalism as Religious Quest, by Thomas R. Dunlap

The Nature of Gold: An Environmental History of the Klondike Gold Rush, by Kathryn Morse

Where Land and Water Meet: A Western Landscape Transformed, by Nancy Langston

The Rhine: An Eco-Biography, 1815–2000, by Mark Cioc

Driven Wild: How the Fight against Automobiles Launched the Modern Wilderness Movement, by Paul S. Sutter

George Perkins Marsh: Prophet of Conservation, by David Lowenthal

Making Salmon: An Environmental History of the Northwest Fisheries Crisis, by Joseph E. Taylor III

Irrigated Eden: The Making of an Agricultural Landscape in the American West, by Mark Fiege

The Dawn of Conservation Diplomacy: U.S.-Canadian Wildlife Protection Treaties in the Progressive Era, by Kirkpatrick Dorsey

Landscapes of Promise: The Oregon Story, 1800–1940, by William G. Robbins

Forest Dreams, Forest Nightmares: The Paradox of Old Growth in the Inland West, by Nancy Langston

The Natural History of Puget Sound Country, by Arthur R. Kruckeberg

CYCLE OF FIRE

Fire: A Brief History, by Stephen J. Pyne

The Ice: A Journey to Antarctica,
by Stephen J. Pyne

Burning Bush: A Fire History of Australia, by Stephen J. Pyne

Fire in America: A Cultural History of Wildland and Rural Fire,
by Stephen J. Pyne

Vestal Fire: An Environmental History, Told through Fire, of Europe and Europe's Encounter with the World,
by Stephen J. Pyne

World Fire: The Culture of Fire on Earth,
by Stephen J. Pyne

Also available:

Awful Splendour: A Fire History of Canada, by Stephen J. Pyne

Lightning Source UK Ltd.
Milton Keynes UK
UKOW04n0315221217
314919UK00009B/492/P

9 780295 999616